Kathrin Daub

Interstitially stabilized cluster complexes of Dy, Ho and Er

Kathrin Daub

Interstitially stabilized cluster complexes of Dy, Ho and Er

New insights into structure and bonding

Südwestdeutscher Verlag für Hochschulschriften

Impressum/Imprint (nur für Deutschland/ only for Germany)
Bibliografische Information der Deutschen Nationalbibliothek: Die Deutsche Nationalbibliothek verzeichnet diese Publikation in der Deutschen Nationalbibliografie; detaillierte bibliografische Daten sind im Internet über http://dnb.d-nb.de abrufbar.

Alle in diesem Buch genannten Marken und Produktnamen unterliegen warenzeichen-, marken- oder patentrechtlichem Schutz bzw. sind Warenzeichen oder eingetragene Warenzeichen der jeweiligen Inhaber. Die Wiedergabe von Marken, Produktnamen, Gebrauchsnamen, Handelsnamen, Warenbezeichnungen u.s.w. in diesem Werk berechtigt auch ohne besondere Kennzeichnung nicht zu der Annahme, dass solche Namen im Sinne der Warenzeichen- und Markenschutzgesetzgebung als frei zu betrachten wären und daher von jedermann benutzt werden dürften.

Verlag: Südwestdeutscher Verlag für Hochschulschriften Aktiengesellschaft & Co. KG
Dudweiler Landstr. 99, 66123 Saarbrücken, Deutschland
Telefon +49 681 37 20 271-1, Telefax +49 681 37 20 271-0
Email: info@svh-verlag.de
Zugl.: Köln, Universität Köln, Diss.,2009

Herstellung in Deutschland:
Schaltungsdienst Lange o.H.G., Berlin
Books on Demand GmbH, Norderstedt
Reha GmbH, Saarbrücken
Amazon Distribution GmbH, Leipzig
ISBN: 978-3-8381-1738-6

Imprint (only for USA, GB)
Bibliographic information published by the Deutsche Nationalbibliothek: The Deutsche Nationalbibliothek lists this publication in the Deutsche Nationalbibliografie; detailed bibliographic data are available in the Internet at http://dnb.d-nb.de.

Any brand names and product names mentioned in this book are subject to trademark, brand or patent protection and are trademarks or registered trademarks of their respective holders. The use of brand names, product names, common names, trade names, product descriptions etc. even without a particular marking in this works is in no way to be construed to mean that such names may be regarded as unrestricted in respect of trademark and brand protection legislation and could thus be used by anyone.

Publisher: Südwestdeutscher Verlag für Hochschulschriften Aktiengesellschaft & Co. KG
Dudweiler Landstr. 99, 66123 Saarbrücken, Germany
Phone +49 681 37 20 271-1, Fax +49 681 37 20 271-0
Email: info@svh-verlag.de

Printed in the U.S.A.
Printed in the U.K. by (see last page)
ISBN: 978-3-8381-1738-6

Copyright © 2010 by the author and Südwestdeutscher Verlag für Hochschulschriften Aktiengesellschaft & Co. KG and licensors
All rights reserved. Saarbrücken 2010

Contents

1. **Introduction**   1

2. **Materials and methods**   5
   - 2.1. Equipment   5
     - 2.1.1. Glove boxes   5
     - 2.1.2. Arc-welding furnace   6
     - 2.1.3. Furnaces   6
     - 2.1.4. Vacuum and inert gas technique   6
     - 2.1.5. Decomposition equipment   6
     - 2.1.6. High vacuum sublimation   6
   - 2.2. Experimental procedure   7
     - 2.2.1. Syntheses of the reactants   9
   - 2.3. Electronic structure calculations   12

3. **Isolated rare-earth clusters**   14
   - 3.1. General aspects of the structure type $M\{ZM_6\}X_{12}$ 14
   - 3.2. Crystal structures of $Ho\{ZHo_6\}I_{12}$ with Z = Fe, Co, Ni, Ir, Pt   16
   - 3.3. Crystal structure of $Dy\{(C_2)Dy_6\}I_{12}$ 24
   - 3.4. Crystal structure of $Dy\{CoDy_{4.53}Y_{1.47}\}I_{12}$ 28
   - 3.5. Electronic structure of $M\{ZM_6\}X_{12}$ type compounds   32

4. **Dimeric rare-earth clusters**   38
   - 4.1. Crystal structure of $\{(C_2)_2Dy_{10}\}Br_{18}$ 38
   - 4.2. Crystal structure of $\{(C_2)_2Er_{10}\}I_{18}$ 42

5. Oligomeric rare-earth clusters — 47

5.1. Crystal structure of $\{Ru_4Ho_{16}\}I_{28}\{Ho_4\}$ 48

5.2. Electronic structure of $\{Ru_4Ho_{16}\}I_{28}\{Ho_4\}$ 56

5.3. Crystal structure of $\{(C_2)_2O_2Dy_{14}\}I_{24}$ 58

6. Cluster chains — 64

6.1. Crystal structure of $\{(C_2)ODy_6\}I_9$ 65

6.2. Crystal structure of $\{IrHo_3\}I_3$ 70

6.3. Electronic structure of $\{IrHo_3\}I_3$ 76

6.4. Crystal structure of $\{(C_2)Er_4\}I_6$ 78

6.5. Electronic structure of $\{(C_2)M_4\}I_6$ type compounds 82

7. Summary and prospects — 84

A. Appendix — 97

1. Introduction

Rare-earth halides exhibiting oxidation numbers of less than three have been known since the first syntheses and characterizations of the products performed by Klemm, Bommer and Döll [1, 2]. As a result, many of them form binary halides with the composition MX_2 (X=Cl, Br, I) which can be divided in two different types: the salt-like and the metallic dihalides. The salt-like dihalides are also called "real" dihalides as they consist of M^{2+} ions with the electron configuration $[Xe]4f^{n+1}5d^0 6s^0$ whereas in metallic dihalides, the rare earth metal has just a formal oxidation number of +2. So the formulation $(M^{3+})(e^-)(I^-)$ is more appropriate, emphasizing the electronic transition from an f- to a d-orbital according to the configuration $[Xe]4f^n 5d^1 6s^0$.

Apart from dihalides, reduced rare earth halides can also be obtained as intermediate phases between the dihalide and trihalide. Thus, the isostructural compounds Dy_5Cl_{11} and Ho_5Cl_{11} crystallize in a fluorite super-structure with additional anions according to the formulation $4\,MCl_2 \cdot MCl_3$ [3, 4]. Even more reduced phases with different compositions such as M_2X_3 (e.g. Gd_2Cl_3) or with a monovalent rare-earth metal as observed in LaI as well as ternary halides are known [1, 5, 6].

By far the greatest structural variety is revealed in the countless cluster compounds of the electron-poor rare-earth metals which have been explored by Corbett, Simon, Meyer and co-workers over the previous two decades [7–13]. Metal clusters are predominantly formed by the early transition metals with relatively large d-orbitals. The cluster formation is realized if the ratio metal/non-metal (e.g. halide) is greater than the preferred coordination number of the metal. Metal-metal bonds are established by excess electrons that are not needed to bind the surrounding ligands. This usually results in a critical situation in the case of the electron-poor

1. Introduction

group three and four metals. However, the electron insufficiency is compensated by introducing an interstitial atom into the cluster center. The interstitial (Z) can be a non-metal like B, C, N, C_2 or a d-metal, mainly from group seven to group nine [14,15], providing additional electrons for M-Z and M-M interactions. The clusters themselves consist of octahedral units and can be classified as isolated (discrete) and condensed clusters. In the case of isolated clusters (Fig. 1.1a), the M_6-unit is peripherically surrounded by halogen atoms, leading to the general composition $\{ZM_6\}X^i_{12}X^a_6$ (with X^i as edge-bridging, inner ligands and X^a as terminal, outer ligands according to Schäfer [16]). Finally, these isolated clusters are connected via halogen atoms to form networks in most cases. If the ratio metal/halide is large, condensed clusters are predominantly obtained, i.e. M_6 clusters sharing common vertices, edges or faces and thus building chains, sheets or networks (Fig. 1.1b).

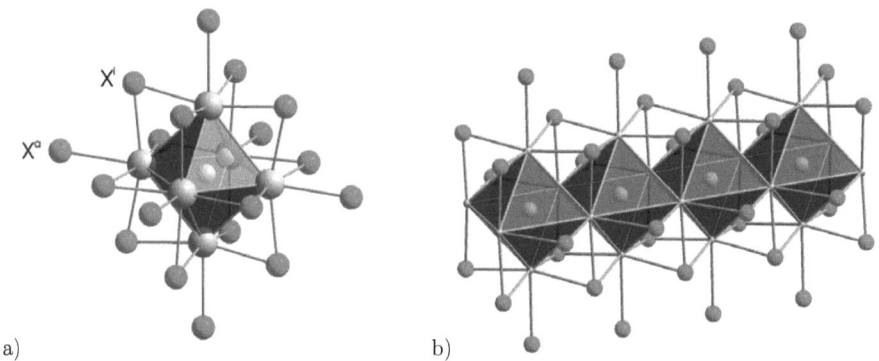

Figure 1.1.: a) $\{ZM_6\}X_{18}$ cluster unit, b) chain of edge-connected M_6 octahedra

But apart from octahedral clusters, tetrahedral, trigonal-bipyramidal, trigonal prismatic. cubic and square-antiprismatic clusters are known as well. As a consequence, there might be an apparent dependency of the cluster size from the size of the interstitial atom.

Status quo

Among the rare-earth halides MX_2, the di-iodides MI_2 (Fig. 1.2) are of great interest: The metallic di-iodides, for instance, have interesting magnetic properties, induced by the delocalization of the $5d^1$ electron. A prominent example is GdI_2 which is not only ferromagnetic, but

1. Introduction

also exhibits a giant magnetoresistivity [17]. But beside the resulting magnetic effects, the $5d^1$ configuration can also lead to cluster formation caused by attractive interactions between those d-orbitals which is realized in the extraordinary PrI_2-V with tetrahedral $\{Pr_4\}$ clusters which are connected to each other according to the formulation $\{Pr_4\}I_4I_{12/3}$. Another feature of PrI_2 is its great variety of observed modifications: Three "metallic" (PrI_2-I, $CuTi_2$type; PrI_2-II, -III, MoS_2 type) as well as a salt-like modification (PrI_2-IV, $CdCl_2$ type) have been investigated in addition to PrI_2-V [18]. The existence of these five modifications indicates a complex system and hence it is challenging to yield pure phases. Despite recent research activities, not all questions have been answered satisfactorily [19].

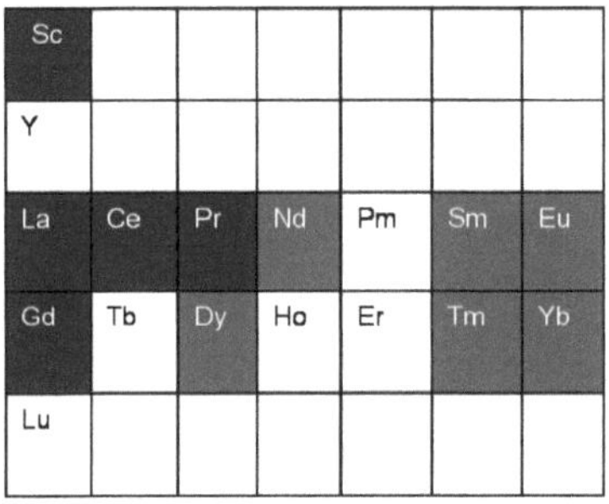

Sc						
Y						
La	Ce	Pr	Nd	Pm	Sm	Eu
Gd	Tb	Dy	Ho	Er	Tm	Yb
Lu						

Figure 1.2.: Rare-earth elements building "metallic" (blue), salt-like (red) or no (white) di-iodides so far.

For the other rare-earth elements, such a distinct variety of modifications has not been known so far. In the case NdI_2, a transition from the salt-like ($SrBr_2$ type) to the metallic modification ($CuTi_2$ type) over an intermediate phase of the CaF_2 type has been observed at high pressures [20]. The transition is thought to be caused by an energetic approximation of the f- and d-orbitals due to the influence of high pressure on the expansion and energy of the f- and d-orbitals and therefore enabling the transition of an electron from the f- into the d-orbital. Moreover, a pressure induced decrease of the cation-cation distance evokes an improved d-orbital overlap and therefore a larger anisotropic conductivity.

A di-iodide of holmium is not known yet, but conproportionation reactions in the system $HoI_3/Ho/C$ led to a phase of the composition $Ho_7I_{12}C$, containing Ho_6 octahedra with an interstitial carbon atom [21]. Together with the condensed cluster compound $[Ho_9C_4O]I_8$, $Ho_7I_{12}C$

has been the only compound consisting of holmium halide clusters. This is in contradiction to the rare-earth elements Sc, Y, La, Pr and Gd revealing quite a vast variety of interstitially stabilized cluster compounds [7, 10, 22, 23].

Also dysprosium halide clusters have not been well investigated until now. Solely two cluster compounds, $\{(C_2)_2Dy_{10}\}I_{18}$ and $\{(C_2)Dy_4\}I_6$, have been reported by Simon and co-workers in 2007 [24]. Other attempts of synthesizing interstitially stabilized clusters only led to the salt-like DyI_2 as the main product [14]. This illustrates that it is not necessarily the magnitude of cluster bonds to achieve successful reactions in the field of solid state chemistry, but also the relative stability of competing phases. Products are hardly predictable, therefore explorative syntheses are commonly applied. As a consequence, surprising and unpredictable, new structure types are not rare at all as it is strikingly demonstrated by the super-tetrahedron $Sc_{24}I_{30}C_{10}$ (Fig. 1.3) [12].

Based on this knowledge, this dissertation is aimed at the synthesis of new holmium, dysprosium and - to some extent - erbium halide cluster complexes and additionally at new insights into the electronic situation in these clusters.

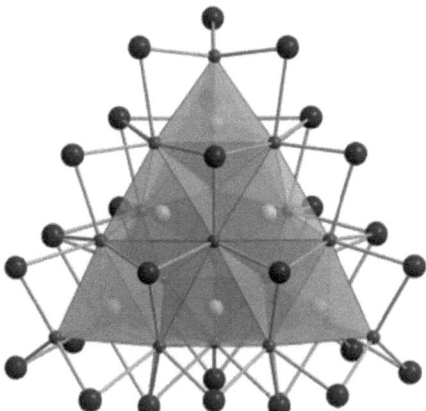

Figure 1.3.: $Sc_{24}I_{30}C_{10}$: A tetrahedron composed of edge-sharing $\{Sc_4C\}$ tetrahedra.

2. Materials and methods

2.1. Equipment

2.1.1. Glove boxes

Because of the air and moisture sensitivity of both, reactants and products, all preparations were carried out in glove boxes (*MBraun, Garching, Germany*) under an argon or nitrogen atmosphere with water and oxygen contents less than 1 ppm. The glove boxes contain an integrated gas purification system that permanently circulates the inert gas and purifies it through a molecular sieve and a copper catalyst. Materials can be transferred into the box through two pressure regulated antechambers. A balance (*Sartorius, Göttingen, Germany*) and a microscope (*type MZ 6, Leica, Wetzlar, Germany*) integrated in the box serve as tools for experimental manipulations.

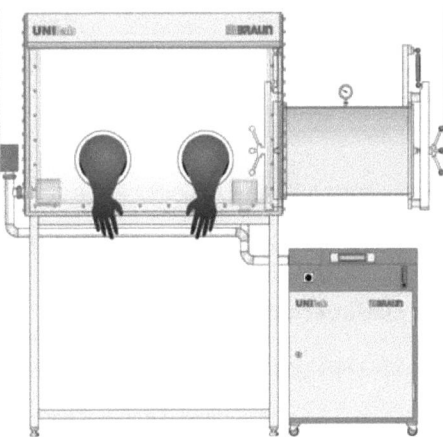

Figure 2.1.: Glove box (*MBraun*).

2.1.2. Arc-welding furnace

Tantalum and niobium containers were exclusively used for the syntheses. After loading the partially welded containers with the reactants, the containers were closed completely by welding it using an arc-welding furnace (*self construction, University of Gießen*). The welding procedure was carried out under a helium atmosphere of 750 mbar and a current of 5-10 A. A tungsten needle and the tantalum/niobium container act as electrodes.

2.1.3. Furnaces

Tube furnaces with chromel/alumel and platinum/platinum-rhodium thermocouples were used for reactions not exceeding a temperature of 950°C. For syntheses with temperatures between 950 and 1200°C muffle furnaces (*Nabertherm, Lilienthal, Germany*) were used.

2.1.4. Vacuum and inert gas technique

The loaded and welded tantalum/niobium containers were encapsulated in silica ampoules under vacuum to avoid oxidation of the metal containers at high temperatures. The silica ampoule was fused at one side and after adding the metal container narrowed at the other side, fixed to the vacuum line by a "quick fit" tube, evacuated and finally entirely fused.

2.1.5. Decomposition equipment

The decomposition equipment consists of a Duran flask (length: 50 cm, diameter 7 cm) and a lid with a gas outlet. The substance is loaded on a silica shuttle and moved into the decomposition apparatus. The gas outlet is connected via a cold trap to a rotary vane pump. The decomposition apparatus is placed into a tube furnace and can undergo various temperature programs.

2.1.6. High vacuum sublimation

The high vacuum equipment (Fig. 2.2) consists of a silica sublimation tube which is put uprightly into a tube furnace and connected to a cold trap and an oil diffusion pump system. The latter contains a rotary vane pump connected upstream to an oil diffusion pump and a

2. Materials and methods

vacuometer. The resulting pressure can be up to 10^{-6} mbar. The substance is filled into a silica tube which is fused at one side and transferred into the sublimation tube. The fused silica tube is coated with another open silica tube. Due to the temperature gradient, the sublimate re-sublimates at the inner walls of the outer silica tube. After having finished the sublimation, the sublimation tube is transferred into a glove box.

Figure 2.2.: Equipment for high vacuum sublimation.

2.2. Experimental procedure

The tantalum and niobium containers were made out of a meter long tantalum tubes which were cut into ca. 4-5 cm long pieces. Afterwards they were cleaned using a solution of conc. H_2SO_4, HNO_3 and HF in the ratio of 2:1:1 in order to remove oxygen impurities. Afterwards they were fused at one side by arc-welding and cleaned again. Then they were transfered into a glove box, filled with the reactants, completely fused and encapsulated in a silica ampoule (Fig. 2.3). The silica ampoules were put in a furnace, exposed to a computer regulated temperature program and eventually opened in a glove box. Suitable single crystals were carefully transferred into a glass capillary and checked concerning their quality by using an X-ray precision camera.

2. Materials and methods

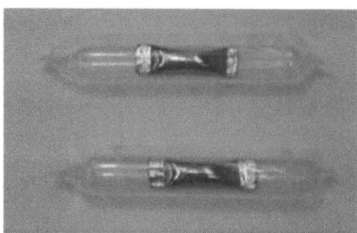

Figure 2.3.: Tantalum containers sealed in silica ampoules.

XRD measurements

Full collections of intensity data of X-ray diffraction of single crystals were carried out by the image plate diffraction systems IPDS I and II (*Stoe, Darmstadt, Germany*). IPDS measurements use image plates containing $BaFBr:Eu^{2+}$ to measure the intensities of the diffracted X-rays. Upon exposure to X-rays, the Eu^{2+} ion is oxidized to Eu^{3+}. The free electrons occupy interstitial lattice sites, forming color centers. Afterwards the image plates are scanned by a He/Ne laser, leading to an activation of the color centers and reducing the Eu^{3+} ions to Eu^{2+} again. As a result, an emission is produced which is measured by a photo-electron multiplier. Then the image plate is radiated with white light to remove possibly still existing color centers.

The analysis of the intensity data was performed with the computer programs SIR-92 [25] and SHELX-97 [26, 27]. A numerical absorption correction was carried out with the programs X-RED [28] and X-SHAPE [29].

Powder diffraction data were recorded on a Huber G 670 instrument (*Huber, Rimsting, Germany*) using $Mo-K_\alpha$ radiation. For the measurements, the samples were filled in glass capillaries with a diameter of 0.3 mm and sealed under inert gas.

Energy dispersive X-ray (EDX) spectroscopy

EDX measurements were recorded on a REM-Zeiss Neon 40 instrument. For the measurements, the sample was adhered to a sample holder in a glove box and rapidly transferred into the instrument, minimizing the exposure to air.

2.2.1. Syntheses of the reactants

Iodides

Lanthanide(III) iodides, MI_3, were synthesized directly from the elements according to the reaction $2\,M + 3\,I_2 \rightarrow 2\,MI_3$. The elements were filled into a silica ampoule and a catalytic amount of ammonium iodide was added.

$$RT \xrightarrow{25°C/h} 100°C \xrightarrow{5°C/h} 150°C \xrightarrow{10°C/h} 185°C\,(14\,h) \xrightarrow{10°C/h} 240°C\,(8\,h) \xrightarrow{20°C/h} RT$$

Afterwards, the ampoule was opened and excessive iodine was sublimated. The product was purified by sublimation under high vacuum and transferred into a glove box.

Bromides and chlorides

Syntheses of the lanthanide(III) bromides and chlorides were carried according to the ammonium halide route by *Meyer* [30]:

$$M_2O_3 + 6\,NH_4X + 6\,HX \rightarrow 2\,(NH_4)_3MX_6 + 3\,H_2O \quad 2\,(NH_4)_3MX_6 \rightarrow 2\,MX_3 + 6\,NH_4X$$

The sesquioxide was dissolved in hydrobromic (hydrochloric) acid and an excess of ammonium bromide (ammonium chloride) was added. The solution was evaporated and the almost dry substance put on a shuttle and transferred into a decomposition apparatus and exposed to the following temperature program:

$$RT \xrightarrow{50°C/h} 120°C(10h) \xrightarrow{50°C/h} 450°C(22h) \xrightarrow{90°C/h} RT$$

Afterwards, the product was purified by sublimation under high vacuum and checked for its purity by powder X-ray diffraction.

2. Materials and methods

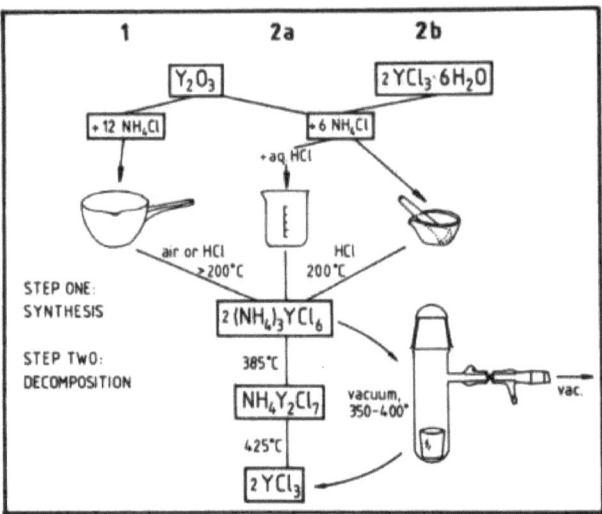

Figure 2.4.: Scheme of the synthesis of lanthanide(III) bromides and chlorides [30].

2. Materials and methods

Applied chemicals

Dysprosium	blocks	Chempur, Karlsruhe, Germany
Dysprosium	powder, ≥99.9 %	Chempur, Karlsruhe, Germany
Holmium	blocks	Chempur, Karlsruhe, Germany
Holmium	powder, ≥99.9 %	Smart Elements, Vienna, Austria
Erbium	blocks	
Yttrium	blocks	Chempur, Karlsruhe, Germany
Yttrium	powder, ≥99.9 %	Chempur, Karlsruhe, Germany
DyI_3	powder	Osram, Munich, Germany
HoI_3	powder	Osram, Munich, Germany
Dy_2O_3	powder, ≥99.9 %	Chempur, Karlsruhe, Germany
Ho_2O_3	powder, ≥99.9 %	Chempur, Karlsruhe, Germany
NH_4Cl	p.a.	Acros, Geel, Belgium
NH_4Br	p.a.	Janssen, Beerse, Belgium
Iodine	p.a.	Acros, Geel, Belgium
HCl	32 %	KMF, Lohmar, Germany
HBr	48 %	Acros, Geel, Belgium
Chromium	powder, ≥99.9 %	Merck, Darmstadt, Germany
Manganese	powder, ≥99.9 %	Chempur, Karlsruhe, Germany
Iron	powder, ≥99.9 %	Merck, Darmstadt, Germany
Cobalt	powder, ≥99.9 %	Merck, Darmstadt, Germany
Nickel	powder, ≥99.9 %	Merck, Darmstadt, Germany
Copper	powder, ≥99.9 %	Merck, Darmstadt, Germany
Zinc	powder, ≥99.9 %	Merck, Darmstadt, Germany
Ruthenium	powder, ≥99.9 %	Merck, Darmstadt, Germany
Rhodium	powder, ≥99.9 %	Merck, Darmstadt, Germany
Osmium	powder, ≥99.9 %	Degussa, Düsseldorf, Germany
Iridium	powder, ≥99.9 %	Chempur, Karlsruhe, Germany
Platinum	powder, ≥99.9 %	Chempur, Karlsruhe, Germany
Graphite	powder, p.a.	Merck, Darmstadt, Germany

Applied computer programs

X-Red [28]	crystal data reduction and correction
WinGX 1.63.02 [31]	graphical interface
SIR-92 [25]	crystal structure solution
SHELXS-97 [26]	crystal structure solution
SHELXL-97 [27]	crystal structure refinement
X-SHAPE [29]	geometrical crystal optimization for absorption correction
PLATON00 [32]	program for checking the space group symmetry
STOE WinXPow [33]	visualization of powder XRD data
X-AREA [34]	integration of single crystal data
DIAMOND 3.0a [35]	program for drawing crystal structures
Origin 7.0 [36]	creation of graphs and diagrams
TB-LMTO-ASA [37]	band structure calculations

2.3. Electronic structure calculations

Electronic structure calculations were carried out with the TB-LMTO-ASA 4.7 program using the atomic spheres approximation and employing the density functional theory (DFT) with the local density approximation for the exchange and correlation energies [37–39]. As a result the intra-atomic Coulomb interactions are neglected which are actually important for systems containing elements with strongly correlating electrons. Therefore the f-electrons of the lanthanide atoms were treated as core electrons in all calculations. Additionally, the local density approximation generally underestimates the band gap in semi-conductors. The following Muffin-Tin Orbitals were used as basis sets:

Ho/Dy: $6s$, $5d$; $6p$ included via the downfolding technique [40]
I: $6s$, $5p$; $5d$ and $4f$ downfolded
Fe/Co/Ni: $4s$, $4p$, $3d$
Ir: $6s$, $6p$, $5d$; $5f$ downfolded
Ru: $5s$, $5p$, $4d$; $4f$ downfolded
Tc: $5s$, $5p$, $4d$, $4f$ downfolded
C: $2s$, $2p$; $3d$ downfolded

2. Materials and methods

The k-points of the first Brillouin zone were chosen via the tetrahedron method [41]. The following numbers of irreducible k-points were used:

M$\{$ZM$_6\}$X$_{12}$ type compounds: 690

$\{$Ru$_4$Ho$_{16}\}$I$_{28}\{$Ho$_4\}$: 451

$\{$IrHo$_3\}$I$_3$: 572

$\{$C$_2$Dy$_4\}$I$_6$: 729

The calculations were interpreted on the basis of DOS curves as well as COHP analyses (crystal orbital Hamilton population) in order to gain information on the bonding situation between atoms [42].

3. Isolated rare-earth clusters

In most cases, rare-earth clusters, either isolated or condensed, derive from M_6X_{12} units, with the halogen atoms bridging all edges of the metal octahedron. Despite their structural similarity, clusters of rare-earth halides are remarkably distinct from binary phases as they encapsulate an interstitial atom, stabilizing the electron-poor metal framework. As a result, the role of metal-interstitial bonding may be more important than that of metal-metal bonding in stabilizing the cluster phase.

In the case of isolated clusters, only the following (ternary) structure types have been synthesized so far: $M\{ZM_6\}X_{12}$ (7-12 type), $\{ZM_6\}X_{10}$ [43], $\{ZM_6\}X_{11}$ [44] and $\{ZM_6\}_2X_{11}$ [45]. Rare earth cluster compounds of the general formula $M\{ZM_6\}X_{12}$, where Z is an interstitial atom, primarily a $3d$ metal or a main group element, have been well explored for the rare earth elements Sc, Y, Pr and Gd [14]. Additionally, compounds of the formula $A_x\{ZR_6\}I_{12+y}$, where A is an alkali metal (Rb or Cs) with x = 1-4 and y = 0-1 and Z = C, C_2, are known for the rare earth elements Pr and Er [46, 47]. In this dissertation, the knowledge of this structure type could be extended to the elements dysprosium and holmium.

3.1. General aspects of the structure type $M\{ZM_6\}X_{12}$

The crystal structure of $M\{ZM_6\}X_{12}$ type compounds consists of crystallographically equivalent metal atoms forming slightly distorted octahedral clusters (D_{3d} symmetry), whose twelve edges are bridged by twelve halogen atoms. Additionally all vertices are saturated with terminal halogen ligands. In general, the halogen ligands can be divided into two groups: Those bridging vertices of one cluster and additionally bonding to non-cluster metal atoms along the c-axis, thus building up almost perfectly octahedral MX_6 entities, and those connecting two clusters among each other by acting as edge-bridging ligands in the one cluster and terminal

3. Isolated rare-earth clusters

ligands in the other (Fig. 3.1). To sum up, the coordination sphere surrounding a metal octahedron consists of six X^i and furthermore six X^{i-a} and X^{a-i} atoms, belonging to two clusters, respectively, and leading to the formulation $M\{ZM_6\}X_6^i X_{6/2}^{i-a} X_{6/2}^{a-i}$.

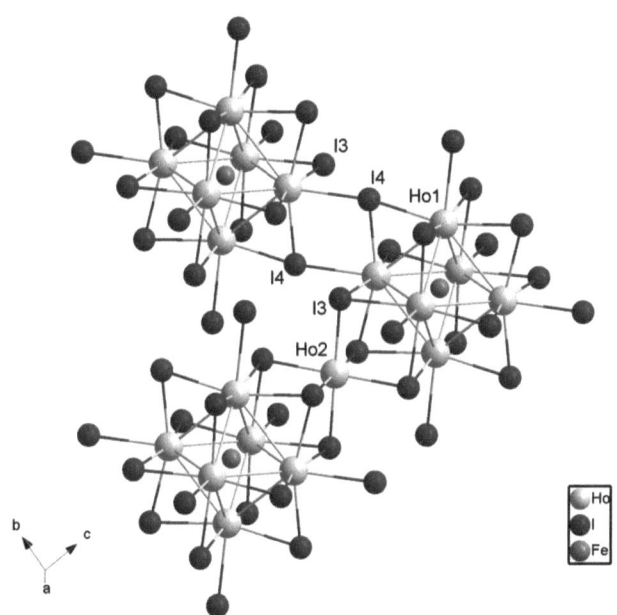

Figure 3.1.: I3 acts as an edge-bridging ligand, additionally bonded to Ho2, whereas I4 connects two clusters among each other. Shown is the example of Ho$\{$FeHo$_6\}$I$_{12}$.

Alternatively, the structure can be described in terms of cubic-closest-packed layers of iodine atoms where every 13th iodine atom is replaced by the interstitial Z. As a result, twelve iodine atoms form a cuboctahedron centered by the interstitial Z and containing the metal atoms in octahedral sites next to the interstitial (Fig. 3.2).

3. Isolated rare-earth clusters

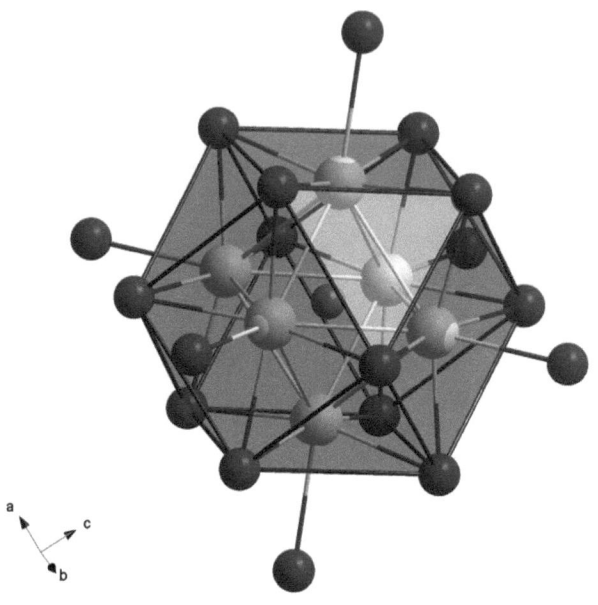

Figure 3.2.: Cuboctahedron built by inner iodine atoms with the interstitial Fe in its center. For the sake of completeness, the terminal iodine atoms are shown as well as the cluster forming metal atoms occupying octahedral sites.

3.2. Crystal structures of $Ho\{ZHo_6\}I_{12}$ with Z = Fe, Co, Ni, Ir, Pt

Holmium compounds of the structure type $M\{ZM_6\}X_{12}$ could be synthesized for the first time. In this regard, single crystals of $Ho\{ZHo_6\}I_{12}$ with Z = Fe, Ni and Pt could be obtained whereas $Ho\{ZHo_6\}I_{12}$ with Z = Co and Ir could be characterized on the basis of powder data.

All compounds crystallize in the space group $R\bar{3}$. Lattice constants and other crystallographic parameters can be found in Tabs. 3.1, 3.3 and 3.5, selected distances are listed in Tabs. 3.2, 3.4 and 3.6. Concerning the powder data, lattice constants could only be determined for

3. Isolated rare-earth clusters

Ho{CoHo$_6$}I$_{12}$.

The cell parameters of Ho{FeHo$_6$}I$_{12}$, Ho{CoHo$_6$}I$_{12}$ and Ho{NiHo$_6$}I$_{12}$ differ remarkably from each other. Even though the cell volumes of Ho{FeHo$_6$}I$_{12}$ (2153.3(5)·10^6 pm^3) and Ho{CoHo$_6$}I$_{12}$ (2151.7(38)·10^6 pm^3) are quite similar, both compounds exhibit strongly differing values for the a- and c-axes, i.e. a = 1529.73(17) pm and c = 1062.52(18) pm for Ho{FeHo$_6$}I$_{12}$ and a = 1545.2(14) pm and c = 1040.7(7) pm for Ho{CoHo$_6$}I$_{12}$, respectively. In contrast, the cell volume of Ho{NiHo$_6$}I$_{12}$ (2179.3(5)·10^6 pm^3) is considerably larger which is contradictory to the analogous praseodymium compounds [14].

In Ho{FeHo$_6$}I$_{12}$ the distances between the cluster metal atoms are 372.97(12) pm and 363.94(11) pm which are shorter than the corresponding ones in Ho{NiHo$_6$}I$_{12}$ (374.73(7) pm and 370.44(7) pm), leading to distances from one apex to the other of 521.10(13) pm in Ho{FeHo$_6$}I$_{12}$ and 526.90(7) pm in Ho{NiHo$_6$}I$_{12}$. Hence, the Ho-Z distances in Ho{FeHo$_6$}I$_{12}$ (260.56(7) pm) are slightly shorter than those in Ho{NiHo$_6$}I$_{12}$ (263.46(4) pm). Despite the larger cluster size, the average distances between the cluster atoms and bridging iodine atoms are a little shorter in Ho{NiHo$_6$}I$_{12}$ (⌀ 311.4 pm) than in the iron analog (⌀ 313.0 pm). The distances between Ho and the terminal iodine atoms are very similar in both compounds, with a value of 329.35(7) pm for the former and 331.16(11) pm for the latter. The main reason for the relatively large unit cell size of the nickel compound is due to the Ho-I distances in the HoI$_6$ entity which are 313.8 pm and thus by far larger than those in the iron analog, exhibiting bond lengths of 301.1 pm (Fig. 3.3).

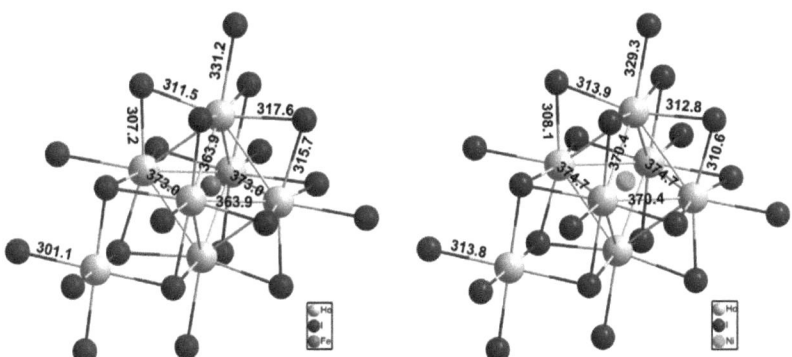

Figure 3.3.: Overview of important interatomic distances (in pm) in Ho{FeHo$_6$}I$_{12}$ (left) and Ho{NiHo$_6$}I$_{12}$ (right).

3. Isolated rare-earth clusters

As expected, the cell size as well as the interatomic distances in the platinum compound are larger than in the $3d$ metal analogs. The values of the a- and c-axes are 1541.0(3) and 1078.33(18) pm, leading to a cell volume of 2217.7(7) $\cdot 10^6$ pm^3. According to that, the Ho-Ho distances in the cluster are 384.95(17) and 381.60(16) pm and larger than those in Ho{NiHo$_6$}I$_{12}$ by about 10 pm. The Ho-I bond lengths are in the same range as the corresponding ones in Ho{FeHo$_6$}I$_{12}$ and Ho{NiHo$_6$}I$_{12}$ (Fig. 3.4).

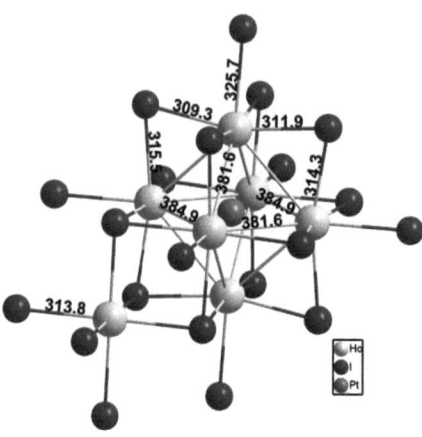

Figure 3.4.: Important interatomic distances in Ho{PtHo$_6$}I$_{12}$.

Even though stoichiometric amounts of all reactants were used, only Ho{CoHo$_6$}I$_{12}$ and Ho{IrHo$_6$}I$_{12}$ could be obtained as pure phases according to X-ray powder diffraction data (Fig. 3.5). Based on its considerably large thermodynamic stability, HoOI is formed as a byproduct once small oxygen impurities are involved in the reaction process. They can either result from impurities in the reactants or more likely from diffusion of oxygen from the silica tube through the walls of the tantalum ampoule which is increased by applying longer reaction times at higher temperatures. Also HoI$_3$ occured as a byproduct, especially in the first attempts of synthesizing Ho{IrHo$_6$}I$_{12}$, but its formation could be suppressed by using an excess of holmium metal in the form of a metal chunk instead of metal powder.

3. Isolated rare-earth clusters

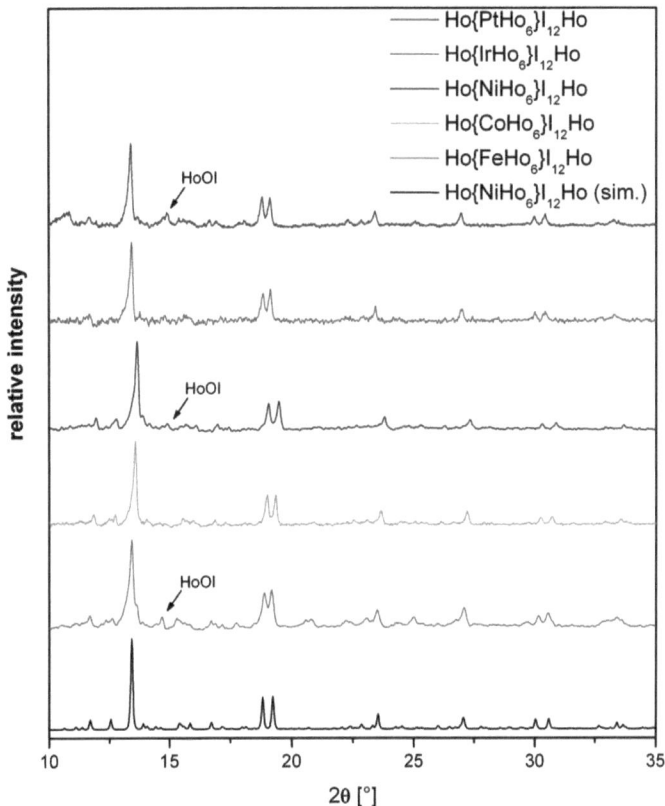

Figure 3.5.: Powder XRD data of Ho{ZHo$_6$}I$_{12}$ with Z = Fe, Co, Ni, Ir, Pt in comparison with the simulated data of Ho{NiHo$_6$}I$_{12}$. Impurities of HoOI are present in the Fe, Ni and Pt species.

Syntheses of $Ho\{ZHo_6\}I_{12}$ with Z = Fe, Co, Ni, Ir, Pt

The compounds were synthesized using 200 mg HoI_3, 46 mg holmium powder and 4 mg iron (cobalt, nickel, 14 mg iridium and platinum) powder. In the case of the iridium analog a ca. 500 mg iridium chunk was used and the metal excess mechanically removed after the reaction was completed. Basically, the stoichiometric amounts were used according to the reaction $4\,HoI_3 + 3\,Ho + Z \rightarrow Ho\{ZHo_6\}I_{12}$. The reaction mixtures were loaded into tantalum containers and exposed to the following temperature program:

$$RT \xrightarrow{50°C/h} 800°C \xrightarrow{15°C/h} 950°C\,(200\,h) \xrightarrow{2°C/h} 680°C \xrightarrow{50°C/h} RT$$

The products consisted of black square prismatic single crystals with edge lengths of 0.1-0.3 mm for the iron, nickel and platinum species and black powder in the case of the cobalt and iridium analogs. Traces of HoOI impurities were detected in the product of the iron, nickel and platinum species whereas $Ho\{CoHo_6\}I_{12}$ and $Ho\{IrHo_6\}I_{12}$ seem to be pure in terms of X-ray powder diffraction (Fig. 3.5).

3. Isolated rare-earth clusters

Table 3.1.: Crystallographic data of Ho{FeHo$_6$}I$_{12}$

Compound	Ho{FeHo$_6$}I$_{12}$
Cell parameters	a = 1529.73(17) pm
	c = 1062.52(16) pm
Cell volume	V = 2153.3(5) ·10^6 pm^3
Formula units Z	3
Crystal system	trigonal
Space group	$R\bar{3}$
Instrument	STOE IPDS I
Radiation	MoK$_\alpha$ (λ = 71.07 pm)
Monochromator	graphite
Temperature	293 K
Density	6.323 g/cm^3
F(000)	3393
Absorption correction	numerical
Absorption coefficient	32.428
Number of measured reflections	6920
Number of independent reflections	1166
Number of parameters	32
R$_{int}$	0.1147
Computing structure solution/refinement	SIR-92, SHELX-97
Scattering factors	International Tables, Vol. C
R$_1$	0.0388 for 861 F$_0$ > 4 σ(F$_0$),
	0.0595 for all data
wR$_2$ (for all data)	0.0960
Goodness of fit S	0.965

Table 3.2.: Selected distances in Ho{FeHo$_6$}I$_{12}$

Atom 1	Atom 2	Distance/pm	Atom 1	Atom 2	Distance/pm
Ho1	Ho1	372.97(12)	Ho1	I4	311.44(11)
Ho1	Ho1	363.94(11)	Ho1	I4	307.22(11)
Ho1	I4	331.16(11)	Ho2	I3	301.06(9)
Ho1	I3	317.58(11)	Ho1	Fe5	260.56(7)
Ho1	I3	315.65(11)			

3. Isolated rare-earth clusters

Table 3.3.: Crystallographic data of Ho{NiHo$_6$}I$_{12}$

Compound	Ho{NiHo$_6$}I$_{12}$
Cell parameters	a = 1530.50(18) pm
	c = 1074.28(17) pm
Cell volume	V = 2179.3(5) ·10^6 pm^3
Formula units Z	3
Crystal system	trigonal
Space group	$R\bar{3}$
Instrument	STOE IPDS I
Radiation	MoK$_\alpha$ (λ = 71.07 pm)
Monochromator	graphite
Temperature	293 K
Density	6.254 g/cm^3
F(000)	3399
Absorption correction	numerical
Absorption coefficient	32.189
Number of measured reflections	9768
Number of independent reflections	1348
Number of parameters	32
R$_{int}$	0.1016
Computing structure solution/refinement	SIR-92, SHELX-97
Scattering factors	International Tables, Vol. C
R$_1$	0.0282 for 1270 F$_0$ > 4 σ(F$_0$),
	0.0308 for all data
wR$_2$ (for all data)	0.0799
Goodness of fit S	1.209

Table 3.4.: Selected distances in Ho{NiHo$_6$}I$_{12}$

Atom 1	Atom 2	Distance/pm	Atom 1	Atom 2	Distance/pm
Ho1	Ho1	374.73(7)	Ho1	I4	312.82(6)
Ho1	Ho1	370.44(7)	Ho1	I4	310.62(6)
Ho1	I5	329.35(7)	Ho1	I5	308.08(6)
Ho1	I5	313.91(7)	Ho1	Ni3	263.46(4)
Ho2	I4	313.71(6)			

Table 3.5.: Crystallographic data of Ho{PtHo$_6$}I$_{12}$

Compound	Ho{PtHo$_6$}I$_{12}$
Cell parameters	a = 1541.0(3) pm
	c = 1078.33(18) pm
Cell volume	V = 2217.7(7) ·10^6 pm^3
Formula units Z	3
Crystal system	trigonal
Space group	$R\bar{3}$
Instrument	STOE IPDS I
Radiation	MoK$_\alpha$ (λ = 71.07 pm)
Monochromator	graphite
Temperature	293 K
Density	6.452 g/cm^3
F(000)	3549
Absorption correction	numerical
Absorption coefficient	35.721
Number of measured reflections	6957
Number of independent reflections	1180
Number of parameters	33
R$_{int}$	0.1400
Computing structure solution/refinement	SIR-92, SHELX-97
Scattering factors	International Tables, Vol. C
R$_1$	0.0505 for 767 F$_0$ > 4 σ(F$_0$),
	0.0906 for all data
wR$_2$ (for all data)	0.1019
Goodness of fit S	0.919

Table 3.6.: Selected distances in Ho{PtHo$_6$}I$_{12}$

Atom 1	Atom 2	Distance/pm	Atom 1	Atom 2	Distance/pm
Ho1	Ho1	384.95(17)	Ho2	I3	313.77(15)
Ho1	Ho1	381.60(16)	Ho1	I3	311.89(18)
Ho1	I4	325.66(18)	Ho1	I4	309.25(19)
Ho1	I4	315.46(18)	Ho1	Pt5	271.02(10)
Ho1	I3	314.27(18)			

3.3. Crystal structure of $\mathrm{Dy}\{(C_2)Dy_6\}I_{12}$

Going to the element next to holmium, i.e. dysprosium, one would expect similar clusters since the atomic radii and physical properties of both elements differ just marginally. In addition, the electronic structure is quite similar, just varying in one f-electron. However, numerous attempts to synthesize analogous dysprosium clusters with endohedral transition metals failed. But the inclusion of carbon as interstitial succeeded instead, resulting in the compound $\mathrm{Dy}\{(C_2)Dy_6\}I_{12}$. As well as the holmium analogs, $\mathrm{Dy}\{(C_2)Dy_6\}I_{12}$ crystallizes in the space group $R\bar{3}$ with the lattice constants a = 1523.3(3) pm, c = 1064.9(3) pm and V = 2140.0(7) $\cdot 10^6$ pm^3. Other crystallographic parameters can be found in Tab. 3.7, selected distances are listed in Tab. 3.8. The most distinctive feature in comparison with other 7-12 cluster complexes is the encapsulation of a C_2 unit in contrast to the common single atom. This leads to an elongation of 92.63 pm along the pseudo fourfold axis of the trigonal anti-prismatic dysprosium cluster (Fig. 3.6).

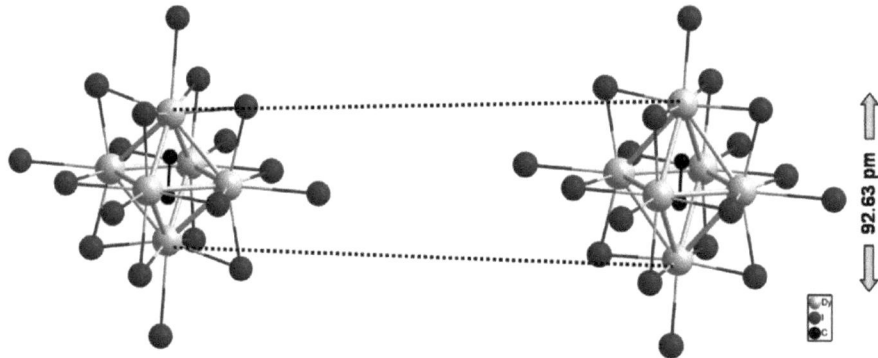

Figure 3.6.: Elongation of the dysprosium cluster along the pseudo fourfold axis in $\mathrm{Dy}\{(C_2)Dy_6\}I_{12}$.

As a consequence, it is more appropriate to refer to the cluster as a trigonal anti-prism instead of a distorted octahedron. The Dy-Dy distances vary between 333.13(6) and 373.06(7) pm. The shortest interactions are those building up the central rectangular plane with values of 333.13(6) and 337.85(5) pm whereas the Dy-Dy interactions along the elongation direction reveal larger values, namely 366.32(6), 368.03(3), 371.38(5) and 373.06(7) pm (Fig. 3.7). The latter are similar to those in $\mathrm{Ho}\{FeHo_6\}I_{12}$, constituting a comparable extension of a C_2 unit

3. Isolated rare-earth clusters

along the bond axis as a $3d$ metal. However, the shorter Dy-Dy distances are due to a minor lateral extension of the C_2 unit due to the small size of a carbon atom. The C-C distance of 149.96(2) pm is slightly shorter than a usual C-C single bond.

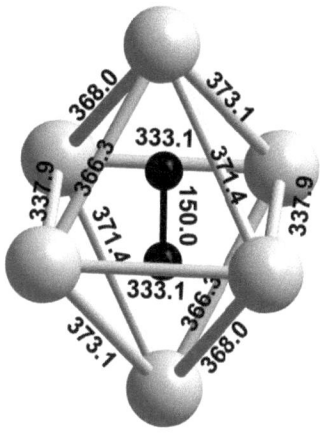

Figure 3.7.: Dy-Dy distances in $Dy\{(C_2)Dy_6\}I_{12}$ (in pm).

Even though condensed rare-earth halide clusters with C or C_2 as interstitials are already known as they can be found in $\{(C_2)Sc_6\}I_{11}$ or $\{(C_2)Sc_4\}I_6$ [44] for instance, $Dy\{(C_2)Dy_6\}I_{12}$ is the first ternary, monomeric compound with an endohedral C_2 dumbbell. Since the C_2 dumbbells can be orientated parallel to each of the three fourfold axes of a regular octahedron, they appear to be statistically disordered in the same way with a site occupation factor of 1/3 for each C atom. As a result, the dysprosium octahedra undergo an elongation parallel to the C_2 units as mentioned above. Consequently, they reveal a corresponding motif of disorder to the interstitials (Fig. 3.8).

On the other hand, neither the DyI_6 units nor the bonded iodine atoms seem to be affected by this disorder as apparent in the quite common displacement ellipsoids Fig. 3.9.

3. Isolated rare-earth clusters

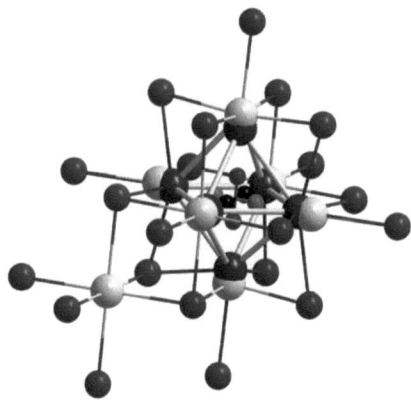

Figure 3.8.: Disordered C_2 dumbbells and Dy atoms in $Dy\{(C_2)Dy_6\}I_{12}$.

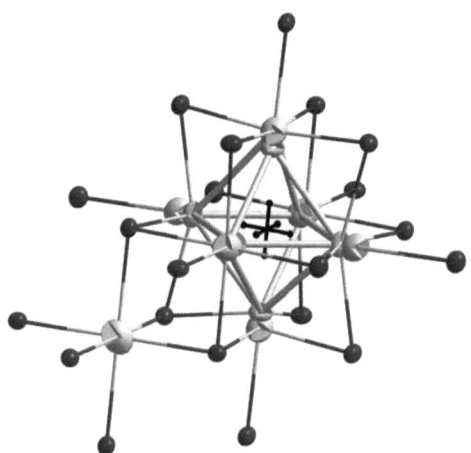

Figure 3.9.: Displacement ellipsoids in $Dy\{(C_2)Dy_6\}I_{12}$ drawn at the 90 % probability level.

Synthesis of $Dy\{(C_2)Dy_6\}I_{12}$

$Dy\{(C_2)Dy_6\}I_{12}$ was synthesized using 200 mg DyI_3, 30 mg dysprosium powder, 6 mg cobalt powder and 4 mg graphite powder. It was actually aimed to get a mixed Co/C_2 dysprosium halide. The reaction mixture was loaded into a tantalum container and exposed to the following

3. Isolated rare-earth clusters

temperature program:

$$\text{RT} \xrightarrow{15°C/h} 1000°C\ (300\ h) \xrightarrow{2°C/h} 600°C \xrightarrow{50°C/h} \text{RT}$$

The product consisted of black square prismatic single crystals with edge lengths of 0.1-0.3 mm which were embedded in a large amount of DyOI.

Table 3.7.: Crystallographic data of Dy$\{(C_2)Dy_6\}I_{12}$

Compound	Dy$\{(C_2)Dy_6\}I_{12}$
Cell parameters	a = 1523.3(3) pm
	c = 1064.9(3) pm
Cell volume	V = 2140.0(7) $\cdot 10^6$ pm^3
Formula units Z	3
Crystal system	trigonal
Space group	$R\bar{3}$
Instrument	STOE IPDS I
Radiation	MoK$_\alpha$ (λ = 71.07 pm)
Monochromator	graphite
Temperature	293 K
Density	6.249 g/cm^3
F(000)	3330
Absorption correction	numerical
Absorption coefficient	31.064
Number of measured reflections	6805
Number of independent reflections	1142
Number of parameters	45
R_{int}	0.0792
Computing structure solution/refinement	SIR-92, SHELX-97
Scattering factors	International Tables, Vol. C
R_1	0.0435 for 868 $F_0 > 4\ \sigma(F_0)$,
	0.0624 for all data
wR_2 (for all data)	0.1088
Goodness of fit S	1.013

Table 3.8.: Selected distances in $Dy\{(C_2)Dy_6\}I_{12}$

Atom 1	Atom 2	Distance/pm	Atom 1	Atom 2	Distance/pm
Dy1A	Dy1B	373.06(7)	Dy3	I4	313.80(4)
Dy1A	Dy1B	371.38(5)	Dy1B	I2	311.93(4)
Dy1A	Dy1B	368.03(6)	Dy1A	I4	310.04(5)
Dy1A	Dy1B	366.32(6)	Dy1B	I2	303.76(4)
Dy1A	Dy1A	337.85(5)	C	C	149.96(2)
Dy1A	Dy1A	333.13(6)			

3.4. Crystal structure of $Dy\{CoDy_{4.53}Y_{1.47}\}I_{12}$

Although dysprosium clusters incorporating transition metals are unknown so far, a mixed Dy/Y cluster could be synthesized. The choice of yttrium as co-reactant is due to the similar atomic radius as dysprosium on the one hand and its notably less electrons on the other hand, making it crystallographically distinguishable from dysprosium. The calculated formula from crystallographic data is $Dy\{CoDy_{4.53}Y_{1.47}\}I_{12}$. The stoichiometric values are well consistent with EDX measurements. As expected, $Dy\{CoDy_{4.53}Y_{1.47}\}I_{12}$ crystallizes in the space group $R\bar{3}$ with the cell parameters a = 1535.80(18)pm, c = 1078.83(18) pm and V = 2203.7(5) $\cdot 10^6$ pm^3. Other crystallographic parameters can be found in Tab. 3.9, selected distances are listed in Tab. 3.10. The cell axes are unusually large compared to those of the holmium 7-12 clusters as dysprosium is just marginally larger than the neighboring holmium. The crystallographic calculations reveal that only the cluster metal atoms are affected by a partial Dy → Y substitution whereas the MI$_6$ units contain exclusively dysprosium atoms. The cluster metal-metal distances range from 375.54(13) to 378.31(10) pm, the cluster metal-iodine distances vary between 309.00(17) and 314.43(12) pm for bridging iodine atoms and 329.51(12) pm for terminal iodine atoms (Fig. 3.10). The Dy-I distances in the DyI$_6$ entities are 315.04(9) pm and thus quite similar to those in $Ho\{NiHo_6\}I_{12}$.

3. Isolated rare-earth clusters

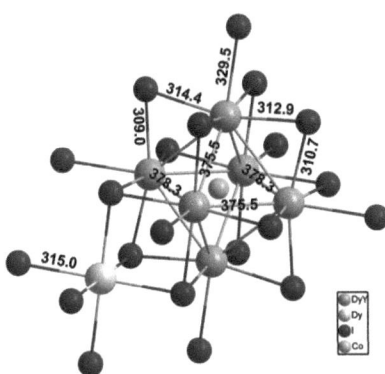

Figure 3.10.: Interatomic distances in Dy{CoDy$_{4.53}$Y$_{1.47}$}I$_{12}$ (in pm).

The reason why it has failed to synthesize dysprosium clusters with transition metals so far could possibly result from electronic circumstances, although Hughbanks and Corbett discuss the responsibility of the competing phase DyI$_2$ for the missing dysprosium clusters with transition metals as interstitials [14]. Even if large amounts of DyI$_2$ accumulate during the attempts of synthesizing such dysprosium clusters, the existence of Dy{(C$_2$)Dy$_6$}I$_{12}$ illustrates that cluster phases can suppress the formation of DyI$_2$. Additionally, it is remarkable that in Dy{CoDy$_{4.53}$Y$_{1.47}$}I$_{12}$, the Y atoms occupy cluster positions instead of MI$_6$ positions, emphasizing that it is probably necessary to stabilize a dysprosium cluster with introducing one to two Y atoms if it encapsulates a transition metal.

Synthesis of Dy{CoDy$_{4.53}$Y$_{1.47}$}I$_{12}$

Dy{CoDy$_{4.53}$Y$_{1.47}$}I$_{12}$ was synthesized using 300 mg DyI$_3$, 45 mg dysprosium powder, 12 mg yttrium powder and 8 mg cobalt powder according to the attempted reaction 4 DyI$_3$ + 2 Dy + Co + Y → Y{CoDy$_6$}I$_{12}$. The reaction mixture was loaded into a tantalum container and exposed to the following temperature program:

$$\text{RT} \xrightarrow{50°C/h} 800°C \xrightarrow{15°C/h} 900°C\ (300\ h) \xrightarrow{2°C/h} 700°C \xrightarrow{50°C/h} \text{RT}$$

The product mostly consisted of a black powder with some black square prismatic crystals.

3. Isolated rare-earth clusters

The crystallographically calculated formula Dy{CoDy$_{4.53}$Y$_{1.47}$}I$_{12}$ was confirmed by EDX measurements.

Another attempt to synthesize a mixed Dy/Y 7-12 type cluster, using the same stoichiometric amounts of the reactants, resulted in a black powder with a 7-12 type compound as main phase and some amounts of DyI$_2$ as well as unreacted DyI$_3$ (Fig. 3.11).

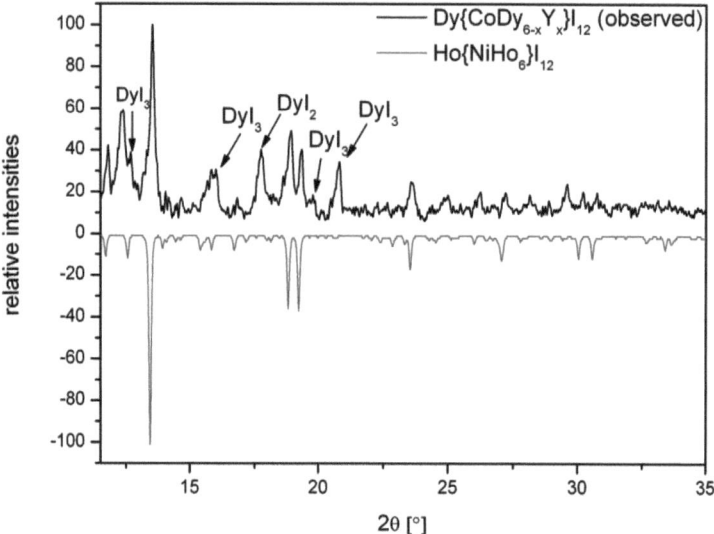

Figure 3.11.: Powder diffraction diagram of a 7-12 type compound containing Dy and/or Y as cluster building rare-earth metals.

3. Isolated rare-earth clusters

Table 3.9.: Crystallographic data of Dy{CoDy$_{4.53}$Y$_{1.47}$}I$_{12}$

Compound	Dy{CoDy$_{4.53}$Y$_{1.47}$}I$_{12}$
Cell parameters	a = 1535.80(18) pm
	c = 1078.83(18) pm
Cell volume	V = 2203.7(5) ·10^6 pm^3
Formula units Z	3
Crystal system	trigonal
Space group	$R\bar{3}$
Instrument	STOE IPDS I
Radiation	MoK$_\alpha$ (λ = 71.07 pm)
Monochromator	graphite
Temperature	293 K
Density	6.214 g/cm^3
F(000)	3414
Absorption correction	numerical
Absorption coefficient	31.357
Number of measured reflections	7067
Number of independent reflections	1194
Number of parameters	35
R$_{int}$	0.1043
Computing structure solution/refinement	SIR-92, SHELX-97
Scattering factors	International Tables, Vol. C
R$_1$	0.0451 for 837 F$_0$ > 4 σ(F$_0$),
	0.0706 for all data
wR$_2$ (for all data)	0.1173
Goodness of fit S	0.944

Table 3.10.: Selected distances in Dy{CoDy$_{4.53}$Y$_{1.47}$}I$_{12}$

Atom 1	Atom 2	Distance/pm	Atom 1	Atom 2	Distance/pm
Dy1|Y1	Dy1|Y1	378.31(10)	Dy1|Y1	I4	312.93(12)
Dy1|Y1	Dy1|Y1	375.54(13)	Dy1|Y1	I4	310.66(12)
Dy1|Y1	I3	329.51(12)	Dy1|Y1	I3	309.00(17)
Dy2	I4	315.04(10)	Co5	Dy1|Y1	266.53(8)
Dy1|Y1	I3	314.43(12)			

3.5. Electronic structure of $M\{ZM_6\}X_{12}$ type compounds

The stabilizing role of the interstitial in bonding within rare-earth metal and zirconium clusters has been investigated before on the basis of extended Hückel molecular orbital (EHMO) calculations [14,22,48–56]. For $3d$ metals the involved orbitals are described in Fig. 3.12. The cluster metal's d-orbitals contribute largely to the strongly bonding orbitals a_{1g} and t_{2g}, interacting with the interstitial's $4s$ and $3d$ orbitals whereas the t_{1u} and a_{2u} have no effect on Z-M bonding and thus just contributing on the M_6I_{12} unit, even though the former shows cluster bonding. The e_g orbital is rather non-bonding and located on the interstitial. 18 electrons are needed to fill the bonding cluster orbitals (above the lower lying iodine and M-I bonding orbitals). This magical number 18 is realized in $Ho\{CoHo_6\}I_{12}$ ($7 \cdot 3 - 12 \cdot 1 + 9 = 18$) for instance. Despite the fact that this number seems to be very important in Zr clusters, rare-earth metal clusters do not strictly obey this 18 electron rule; examples with fewer or even more electrons (like in $Ho\{NiHo_6\}I_{12}$) are widely known. This eminent difference between zirconium and rare-earth metal clusters could be due to the higher lying atomic orbitals of the more electropositive rare-earth elements [14].

The situation for compounds with a main group element as interstitial is somewhat different. The s and p valence orbitals of the interstitial interact with the cluster's a_{1g} and t_{1u} levels and lower them accordingly. The corresponding a_{1g}^* and t_{1u}^* levels are very high-lying and thus not capable of electron accommodation. As a result, the HOMO – LUMO separation is constituted by the t_{2g} level as the HOMO and the a_{2u} level as the LUMO, adding up 14 electrons to the HOMO – LUMO boundary.

3. Isolated rare-earth clusters

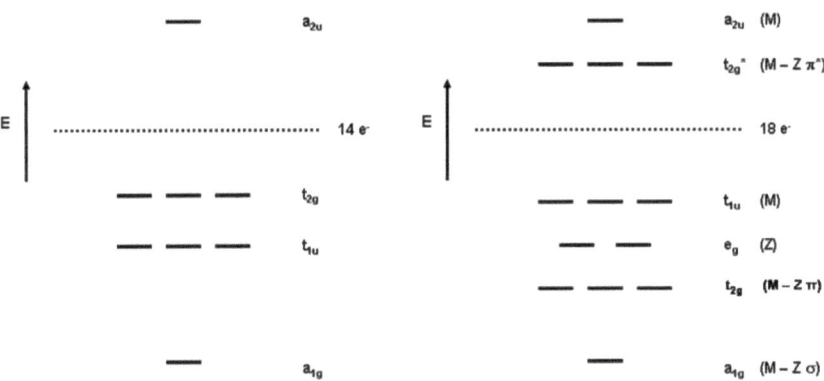

Figure 3.12.: Relative energy levels of molecular orbitals of M_6X_{12} clusters with interstitial main group element [22] (left) and transition metal [14] (right).

Despite the long history of 7-12 rare-earth cluster compounds, calculations of the extended structure on the basis of DOS and COHP curves have not been performed. In Figs. 3.13 and 3.14 the DOS and COHP curves of Ho$\{$CoHo$_6\}$I$_{12}$ are shown.

3. Isolated rare-earth clusters

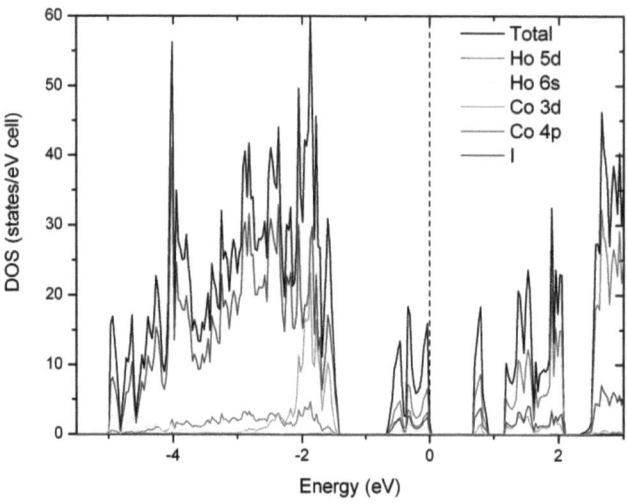

Figure 3.13.: Projected DOS of Ho{CoHo$_6$}I$_{12}$.

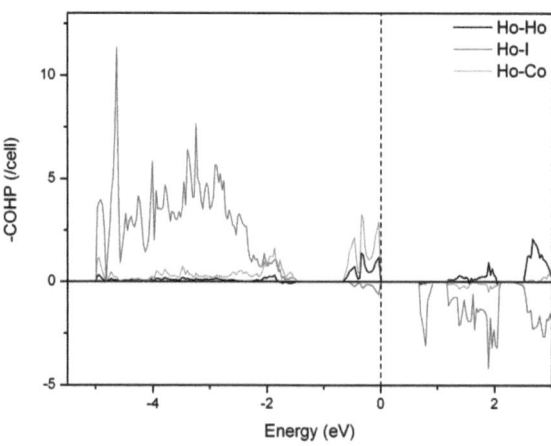

Figure 3.14.: COHP curves of Ho{CoHo$_6$}I$_{12}$.

3. Isolated rare-earth clusters

The COHP curves are of particular interest concerning the bonding situation. Beside the low-lying, rather ionic, bonding Ho-I interactions, bonding Ho-Co interactions occur below the Fermi level. In contrast to that, there are only small bonding interactions between Ho and Ho below the Fermi level, but additional ones above the Fermi level, i.e. representing empty states whereas all bonding Ho-Co states are below the Fermi level. This leads to the conclusion that the cluster formation is predominantly assigned by Ho-Co, or in general Ho-Z, instead of Ho-Ho bonding. On the basis of the projected DOS curves, the character of the corresponding orbitals that interact can be determined. The Co $3d$ states are located at energies around -2 eV and responsible for the Ho-Co interactions in this region. The stronger Ho-Co interactions right below the Fermi level are widely formed by Co $4p$ and Ho $5d$ orbitals.

The Fermi level is located just above the bonding Ho-Ho and Ho-Co and anti-bonding Ho-I interactions beginning at -1 eV and followed by a band gap of 0.7 eV constituting a semiconductor. According to the EHMO calculations mentioned above, the "magical" electron number of 18 is realized in Ho$\{$CoHo$_6\}$I$_{12}$. And also in the extended structure calculations, the electronic situation in Ho$\{$CoHo$_6\}$I$_{12}$ is particularly interesting as the Fermi level exactly separates the bonding Ho-Co interactions from the corresponding anti-bonding ones. The situations in the electron-poorer Ho$\{$FeHo$_6\}$I$_{12}$ and richer Ho$\{$NiHo$_6\}$I$_{12}$ are shown in Figs. 3.15 and 3.16. The Fermi level in Ho$\{$FeHo$_6\}$I$_{12}$ is shifted towards the bonding Ho-Ho and Ho-Fe as well as slightly anti-bonding Ho-I states, i.e. in contrast to Ho$\{$CoHo$_6\}$I$_{12}$, it is not possible to fill all the bonding Ho-Z states. The situation in Ho$\{$NiHo$_6\}$I$_{12}$ is exactly the opposite: The higher electron number shifts the Fermi level towards anti-bonding Ho-I states, i.e. leading to a somewhat more unstable structure. These facts resemble the results of the EHMO calculations.

3. Isolated rare-earth clusters

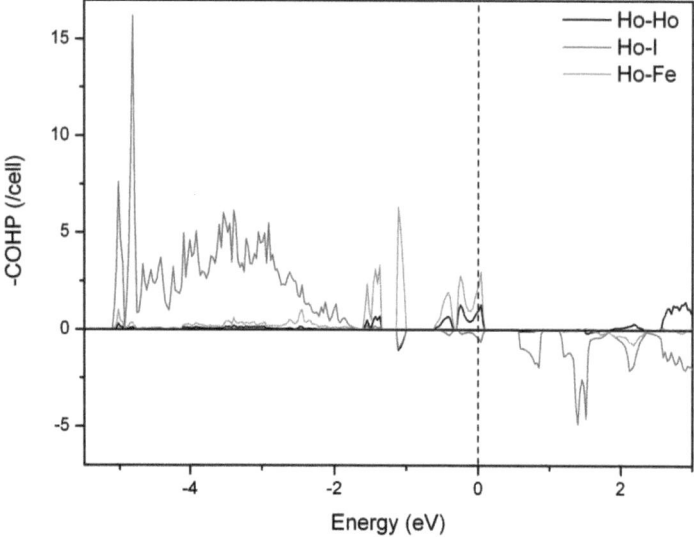

Figure 3.15.: COHP curves of Ho{FeHo$_6$}I$_{12}$.

The only illogicality in the latter compounds is that the Fermi levels in the corresponding DOS curves lie at DOS being unequal to zero, i.e. those compounds actually must be conductors which is not true for monomeric compounds like 7-12 type clusters. This can be due to the uneven electron number in these compounds. Therefore spin-polarized calculations have been performed for Ho{NiHo$_6$}I$_{12}$. The result is shown in Fig 3.17 and reveals a band gap at the Fermi level for both, the α and β electrons.

3. Isolated rare-earth clusters

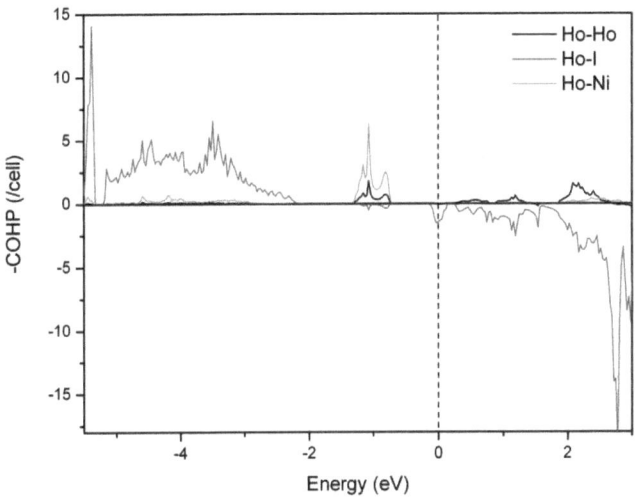

Figure 3.16.: COHP curves of Ho{NiHo$_6$}I$_{12}$.

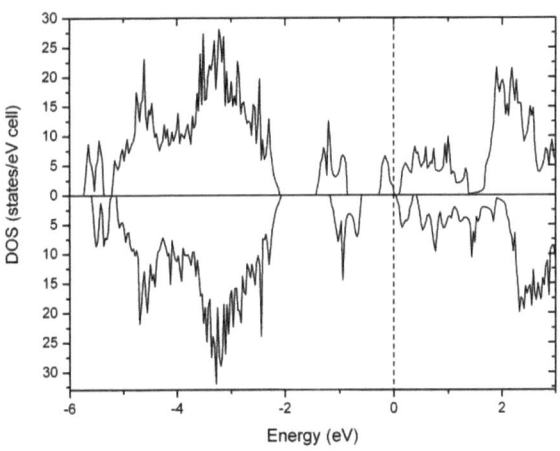

Figure 3.17.: Spin-polarized DOS curves of Ho{NiHo$_6$}I$_{12}$.

4. Dimeric rare-earth clusters

The first step of cluster condensation leads to bi-octahedral units as realized in the closely related structure types $\{(C_2)_2M_{10}\}X_{18}$, $\{(C_2)_2M_{10}\}X_{17}$ and $\{(C_2)_2M_{10}\}X_{16}$. The structural motif consists of two metal octahedra condensed via a common edge forming dimeric clusters. Both octahedra are centered by a C_2 unit with the free edges bridged by halogen atoms as it is present in the M_6X_{12}-type cluster complexes and the vertices bonded to terminal halogen atoms. In this dissertation, two new $\{(C_2)_2M_{10}\}X_{18}$-type compounds have been synthesized, namely $\{(C_2)_2Dy_{10}\}Br_{18}$ and $\{(C_2)_2Er_{10}\}I_{18}$, as well as $\{(C_2)_2Dy_{10}\}I_{18}$, recently published by Simon and co-workers [24].

4.1. Crystal structure of $\{(C_2)_2Dy_{10}\}Br_{18}$

$\{(C_2)_2Dy_{10}\}Br_{18}$ crystallizes in the monoclinic space group $P2_1/c$ with the cell parameters a = 973.99(12) pm, b = 1633.98(15) pm, c = 1324.69(19) pm, β = 120.87(1)° and V = 1809.6(4) $\cdot 10^6$ pm^3. Other crystallographic parameters can be found in Tab. 4.1, selected distances are listed in Tab. 4.2. As in Dy$\{(C_2)Dy_6\}I_{12}$, the C_2 units cause a tetragonal distortion of the octahedron. Therefore, the Dy-Dy distances are quite large along this direction (363.09(7) pm - 403.69(12) pm) whereas the distances perpendicular to the elongation are considerably smaller. In addition, the latter can be distinguished in Dy-Dy bonds forming common edges and those just bridged by bromine atoms. The former are remarkably smaller, exhibiting a bond length of 318.32(11) pm in contrast to the latter with values of 349.38(8) - 355.58(9) pm (Fig. 4.1). Another striking fact is that both octahedra are slightly tilted away from each other (Fig. 4.2), probably to minimize repulsion between bromide ligands. This results in the larger Dy-Dy distances close to the center of the dimer (Fig. 4.2). The distance between two neighboring, apical dysprosium atoms of each octahedron is 419.33(9) pm, hence no real interaction is established

4. Dimeric rare-earth clusters

between these atoms. The C-C distance is 143.7(13) pm, approximately corresponding to the values in other $\{(C_2)_2M_{10}\}X_{18}$-type compounds [22].

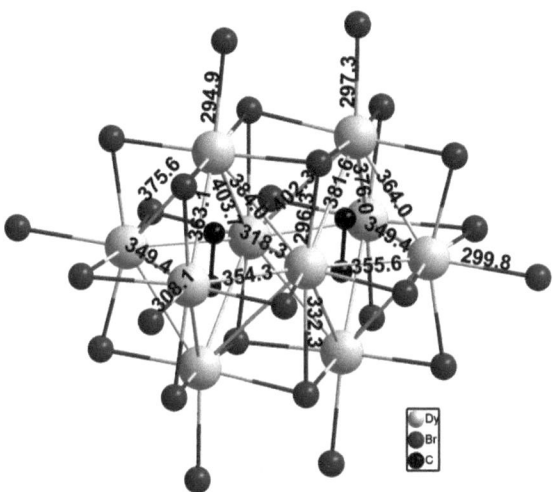

Figure 4.1.: Dy-Dy and selected Dy-Br distances in $\{(C_2)_2Dy_{10}\}Br_{18}$ (in pm).

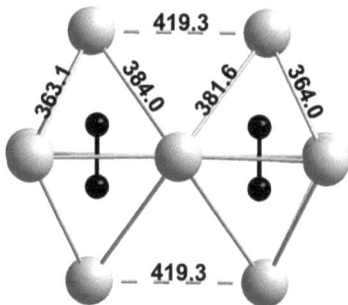

Figure 4.2.: Lateral view of a dimeric dysprosium cluster in $\{(C_2)_2Dy_{10}\}Br_{18}$ (distances in pm).

The metal atoms are surrounded by bromine atoms above the cluster edges and vertices according to the formulation $\{(C_2)_2Dy_{10}\}Br^i_{10}Br^{i-a}_{8/2}Br^{a-i}_{8/2}$, i.e. some of them belong to a single cluster unit while others connect neighboring cluster units, leading to a three-dimensional

4. Dimeric rare-earth clusters

network (Fig. 4.3). In contrast to $\{(C_2)_2Gd_{10}\}Cl_{17}$ [57], $\{(C_2)_2Gd_{10}\}Cl_{16}$ or $\{(C_2)_2Gd_{10}\}Cl_{15}$ [58], in which neighboring clusters are additionally aligned via i-i contacts, $\{(C_2)_2Dy_{10}\}Br_{18}$ only reveals i-a and a-i contacts, respectively. The structure consists of nine crystallographically inequivalent bromine atoms. Four of them are terminal ligands with Dy-Br distances between 294.88(12) and 308.12(12) pm (Fig. 4.1). It is notable that the Dy-Br distances parallel to the elongation are the shortest ones. The distances to the edge-capping bromine atoms range from 271.57(15) to 293.40(12) pm. Additionally, the clusters are bridged by another type of bromine atom, connecting the apical dysprosium atoms as well as one of the metal atoms of the common edge of the condensed clusters. The latter Dy-Br bonds exhibit relatively large bond lengths of 296.30(12)) and 332.21(12) pm.

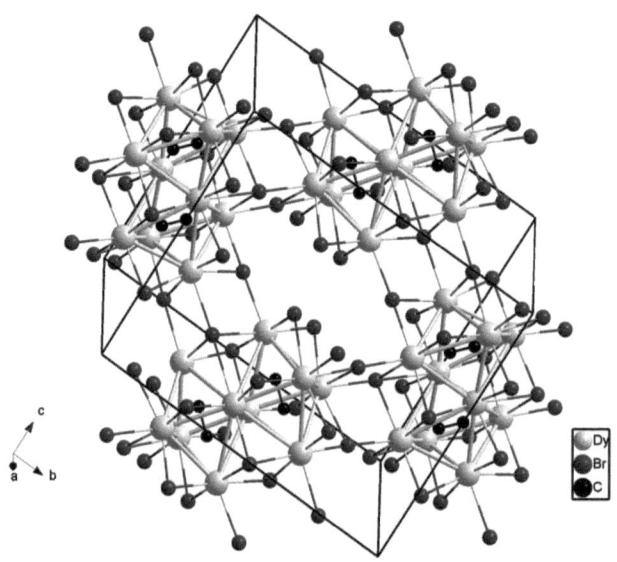

Figure 4.3.: Unit cell of $\{(C_2)_2Dy_{10}\}Br_{18}$.

Synthesis of $\{(C_2)_2Dy_{10}\}Br_{18}$

$\{(C_2)_2Dy_{10}\}Br_{18}$ was synthesized using 150 mg $DyBr_3$, 85 mg dysprosium powder and 8 mg graphite powder, actually in order to synthesize a more condensed phase according to

4. Dimeric rare-earth clusters

5 DyBr$_3$ + 7 Dy + C → 3 Dy$_4$Br$_5$C. The reaction mixture was loaded into a tantalum container and exposed to the following temperature program:

$$\text{RT} \xrightarrow{50°\text{C/h}} 800°\text{C} \xrightarrow{15°\text{C/h}} 1000°\text{C} \ (200\ \text{h}) \xrightarrow{5°\text{C/h}} 650°\text{C} \xrightarrow{50°\text{C/h}} \text{RT}$$

The product consisted of black, cuboid crystals with edge lengths of ca. 0.2 mm and a huge amount of DyOBr impurities.

Table 4.1.: Crystallographic data of $\{(C_2)_2Dy_{10}\}Br_{18}$

Compound	$\{(C_2)_2Dy_{10}\}Br_{18}$
Cell parameters	a = 973.99(12) pm
	b = 1633.98(15) pm
	c = 1324.69(19) pm
	β = 120.869(9)°
Cell volume	V = 1809.6(4) ·10^6 pm^3
Formula units Z	2
Crystal system	monoclinic
Space group	$P2_1/c$
Instrument	STOE IPDS I
Radiation	MoK$_\alpha$ (λ = 71.07 pm)
Monochromator	graphite
Temperature	293 K
Density	5.710 g/cm^3
F(000)	2628
Absorption correction	numerical
Absorption coefficient	40.236
Number of measured reflections	23768
Number of independent reflections	3935
Number of parameters	136
R$_{int}$	0.0961
Computing structure solution/refinement	SIR-92, SHELX-97
Scattering factors	International Tables, Vol. C
R$_1$	0.0326 for 2989 F$_0$ > 4 σ(F$_0$),
	0.0498 for all data
wR$_2$ (for all data)	0.0785
Goodness of fit S	1.010

Table 4.2.: Selected distances in $\{(C_2)_2Dy_{10}\}Br_{18}$

Atom 1	Atom 2	Distance/pm	Atom 1	Atom 2	Distance/pm
Dy1	Dy4	381.63(7)	Dy3	Br15	293.40(12)
Dy3	Dy1	375.95(9)	Dy2	Br12	292.58(13)
Dy6	Dy2	375.60(8)	Dy2	Br14	288.27(12)
Dy1	Dy2	363.99(8)	Dy3	Br14	287.59(14)
Dy3	Dy6	363.09(7)	Dy6	Br13	286.55(14)
Dy2	Dy4	355.58(9)	Dy4	Br11	285.91(12)
Dy3	Dy4	354.34(8)	Dy4	Br19	285.64(14)
Dy3	Dy2	349.38(8)	Dy1	Br13	285.39(12)
Dy4	Br20	332.31(12)	Dy1	Br15	284.73(12)
Dy4	Dy4	318.32(11)	Dy6	Br12	284.13(14)
Dy3	Br20	308.12(12)	Dy2	Br17	283.88(13)
Dy2	Br12	299.78(14)	Dy3	Br16	283.54(13)
Dy1	Br14	297.30(12)	Dy6	Br16	274.85(12)
Dy4	Br13	296.30(12)	Dy1	Br17	274.13(15)
Dy6	Br15	294.88(12)	Dy2	Br19	272.13(12)
Dy1	Br20	294.34(14)	Dy3	Br11	271.57(15)
Dy6	Br20	294.32(12)	C1	C2	143.7(13)

4.2. Crystal structure of $\{(C_2)_2Er_{10}\}I_{18}$

$\{(C_2)_2Er_{10}\}I_{18}$ crystallizes in the space group $P2_1/n$ and is isotypic with $\{(C_2)_2Dy_{10}\}Br_{18}$, but the cell parameters are larger due to the size of the iodine atoms, with a = 1045.01(13) pm, b = 1707.22(17) pm, c = 1237.10(16) pm, β = 105.110(15)° and V = 2130.8(4) ·10^6 pm^3. Other crystallographic parameters can be found in Tab. 4.3, selected distances are listed in Tab. 4.4.

The Er-Er distances along the tetragonal distortion range from 364.25(12) to 400.54(13) pm. As well as in $\{(C_2)_2Dy_{10}\}Br_{18}$, Er-Er bonds forming common edges (320.54(17) pm) are smaller than those being bridged by iodine atoms (347.37(12) - 362.05(12) pm). The distance between two apical erbium atoms is 426.78(12) pm and thus a little larger than the corresponding distance in $\{(C_2)_2Dy_{10}\}Br_{18}$.

4. Dimeric rare-earth clusters

The Er-I distances are similar to those in the $\{(C_2)_2Dy_{10}\}I_{18}$ analog [24]. The four crystallographically inequivalent terminal ligands exhibit Er-I distances between 320.55(15) and 346.45(15) pm. The distances to the edge-capping iodine atoms range from 290.38(15) to 314.50(16) pm. An overview of all interatomic distances is illustrated in Fig. 4.4.

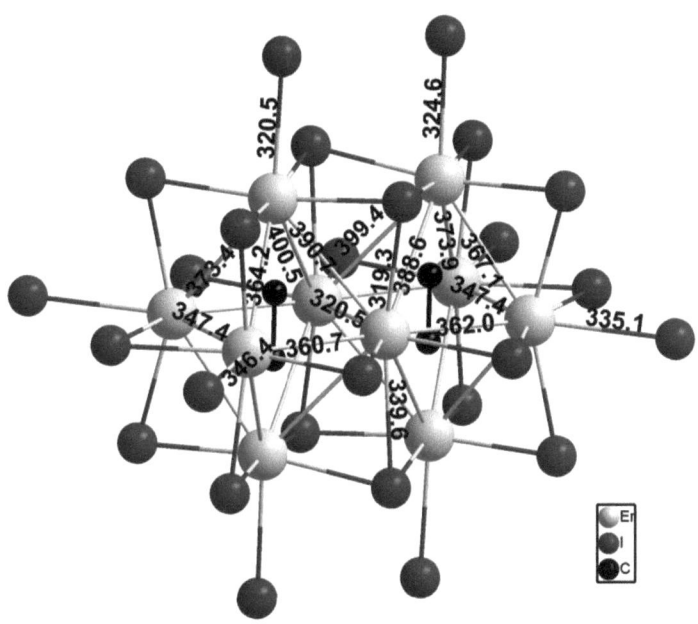

Figure 4.4.: Interatomic distances in $\{(C_2)_2Er_{10}\}I_{18}$ (in pm).

Synthesis of $\{(C_2)_2Er_{10}\}I_{18}$

$\{(C_2)_2Er_{10}\}I_{18}$ was synthesized using 200 mg ErI_3, 85 mg erbium powder and an excess of 8 mg graphite powder, actually in order to synthesize a more condensed phase according to $5\,ErI_3 + 7\,Er + 3\,C \rightarrow 3\,Er_4I_5C$. The reaction mixture was loaded into a tantalum container and exposed to the following temperature program:

$$RT \xrightarrow{50°C/h} 700°C \xrightarrow{10°C/h} 1000°C\ (200\ h) \xrightarrow{2°C/h} 800°C \xrightarrow{5°C/h} 700°C \xrightarrow{50°C/h} RT$$

4. Dimeric rare-earth clusters

The product consisted of a mixture of black, cuboid crystals as well as some black needles of the composition $\{(C_2)Er_4\}I_6$ and some ErOI impurities. The powder diffraction diagram shows that $\{(C_2)_2Er_{10}\}I_{18}$ is the main phase though.

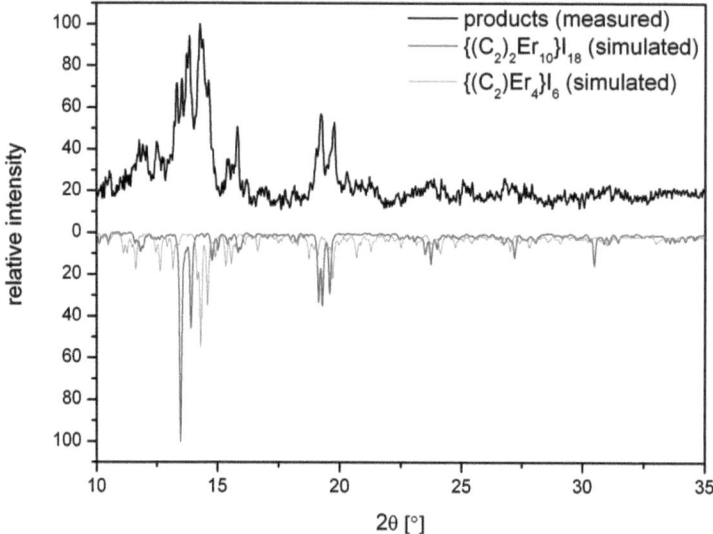

Figure 4.5.: Powder diffraction analysis of the product mixture. $\{(C_2)_2Er_{10}\}I_{18}$ forms the main phase, but the existence of $\{(C_2)Er_4\}I_6$ is evident.

4. Dimeric rare-earth clusters

Table 4.3.: Crystallographic data of $\{(C_2)_2Er_{10}\}I_{18}$

Compound	$\{(C_2)_2Er_{10}\}I_{18}$
Cell parameters	a = 1045.01(13) pm
	b = 1707.22(17) pm
	c = 1237.10(16) pm
	$\beta = 105.110(15)°$
Cell volume	V = 2130.8(4) $\cdot 10^6$ pm^3
Formula units Z	2
Crystal system	monoclinic
Space group	$P2_1/n$
Instrument	STOE IPDS I
Radiation	MoK$_\alpha$ (λ = 71.07 pm)
Monochromator	graphite
Temperature	293 K
Density	6.242 g/cm^3
F(000)	3316
Absorption correction	numerical
Absorption coefficient	32.494
Number of measured reflections	16544
Number of independent reflections	5149
Number of parameters	135
R_{int}	0.1865
Computing structure solution/refinement	SIR-92, SHELX-97
Scattering factors	International Tables, Vol. C
R_1	0.0558 for 3534 $F_0 > 4\ \sigma(F_0)$,
	0.0820 for all data
wR_2 (for all data)	0.1405
Goodness of fit S	0.898

Table 4.4.: Selected distances in $\{(C_2)_2Er_{10}\}I_{18}$

Atom 1	Atom 2	Distance/pm	Atom 1	Atom 2	Distance/pm
Er1	Er5	400.54(13)	Er5	I7	313.11(15)
Er1	Er5	390.75(12)	Er4	I14	311.65(15)
Er2	Er5	388.57(12)	Er3	I13	310.88(15)
Er2	Er4	373.93(13)	Er1	I12	308.64(15)
Er1	Er3	373.44(12)	Er3	I9	308.05(15)
Er2	Er3	367.08(12)	Er4	I9	305.91(14)
Er1	Er4	364.25(12)	Er3	I11	305.24(16)
Er3	Er5	362.05(12)	Er4	I8	304.28(15)
Er4	Er5	360.73(12)	Er1	I6	303.81(14)
Er3	Er4	347.37(12)	Er2	I6	302.53(15)
Er4	I12	346.45(15)	Er2	I14	300.81(15)
Er5	I12	339.57(15)	Er1	I13	300.51(14)
Er3	I13	335.07(14)	Er1	I8	293.46(15)
Er2	I9	324.57(15)	Er2	I11	293.16(14)
Er1	I14	320.55(15)	Er3	I7	291.09(16)
Er5	Er5	320.54(17)	Er4	I10	290.38(15)
Er5	I6	319.27(15)	C1	C2	148(2)
Er5	I10	314.50(16)			

5. Oligomeric rare-earth clusters

The next step of condensation would be a sort of triple octahedron which has not been known until the syntheses of $La_{14}I_{20}(C_2)_3$ [59] and $\{IrGd_{11}\}Br_{15}$ in 2009 [60]. Except for this new structure type involving trimers, tetrameric clusters are quite common, constituting a pool of $\{Z_4M_{16}\}$-type clusters. These clusters are based on oligomeric cluster units, consisting of four condensed octahedra which encapsulate transition metal atoms. These transition metals are arranged in the form of a tetrahedron (Fig. 5.1). An overview of compounds containing oligomeric $\{Z_4M_{16}\}I_{36}$-type cluster complexes is given in Tab. 5.1.

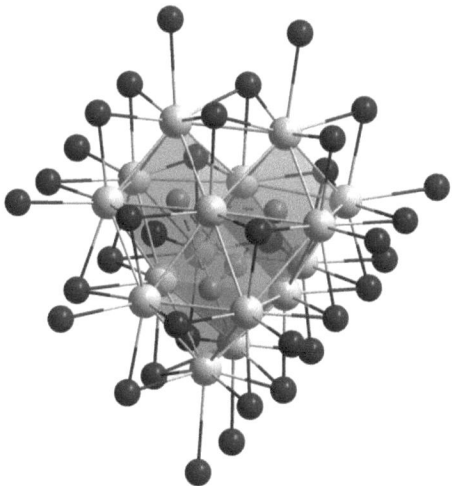

Figure 5.1.: $\{Z_4M_{16}\}I_{36}$ unit. The tetrahedral arrangement of the interstitials is emphasized by fragmented lines.

5. Oligomeric rare-earth clusters

Table 5.1.: $\{Z_4M_{16}\}I_{36}$-type clusters

type	compound	space group	electrons
1	$\{Fe_4Sc_{16}\}Cl_{20}$ [61]	$P4_2/nmm$	15
	$\{(Fe,Os)_4Sc_{16}\}Br_{20}$ [61, 62]		15
	$\{Ru_4Y_{16}\}(Br,I)_{20}$		15
2	$\{Ir_4Y_{16}\}Br_{24}$ [62, 63]	$Fddd$	15
3	$\{(Ru,Os,Ir)_4Sc_{16}\}Cl_{24}\cdot 4ScCl_3$ [64]	$I4_1/a$	14/15
	$\{(Ru,Ir)_4Y_{16}\}Br_{24}\cdot 4YBr_3$ [62]		14/15
4	$\{(Fe,Os,Mn)_4Sc_{16}\}Br_{28}\{Sc_4\}$ [65]	$P\bar{4}3m$	15+1/14+1
	$\{Mn_4Gd_{16}\}I_{28}\{Gd_4\}$ [66]		15
	$\{Ru_4Ho_{16}\}I_{28}\{Ho_4\}$		16

The oligomers exhibit almost tetrahedral symmetry, but packing effects and halide bridging can cause symmetry reduction to D_{2d}, D_2 or S_4 symmetry, respectively. The stability of these tetrameric building blocks is remarkable as they occur in a variety of structure types and incorporate a number of different transition metals. The tetramers belonging to the type 1 structure are surrounded by 36 halogen atoms, with 32 of them connecting eight neighboring cluster units. The type 2 structure is realized by adding four halogen atoms followed by a slight change in the connecting mode between the oligomers. In contrast to that, the type 3 and 4 structures exhibit additional building blocks, namely MX_3 units (type 3), which are chain-like aligned and connect oligomeric clusters among each other, and M_4X_8 tetrahedra (type 4) representing an extra cluster unit besides the oligomeric $\{Z_4M_{16}\}$ unit.

5.1. Crystal structure of $\{Ru_4Ho_{16}\}I_{28}\{Ho_4\}$

$\{Ru_4Ho_{16}\}I_{28}\{Ho_4\}$ crytallizes in the cubic space group $P\bar{4}3m$ (type 4 structure) with the cell parameters a = 1221.14(14) pm and V = 1820.9(4) $\cdot 10^6$ pm^3. Other crystallographic parameters can be found in Tab. 5.2, selected distances are listed in Tab. 5.3.

As a type 4 oligomer, $\{Ru_4Ho_{16}\}I_{28}\{Ho_4\}$ consists of a $\{Ru_4Ho_{16}\}I_{36}$ tetramer with ideal T_d symmetry. There are two possible ways of viewing at the tetrameric building unit $\{Z_4M_{16}\}$: On

5. Oligomeric rare-earth clusters

the one hand the clusters can be described in terms of a condensation product of 2 + 2 $\{ZM_6\}$ octahedra connected via three common edges among each other. This leads to a distortion of every single octahedron and a shift of the interstitial ruthenium towards the center of the holmium oligomer, thus forming a Ru tetrahedron with Ru-Ru distances of 345.81(3) pm (Fig. 5.2). As a consequence, the interactions between ruthenium atoms are rather weak compared to the shortest Ru-Ru distance in metallic ruthenium (265 pm) [67].

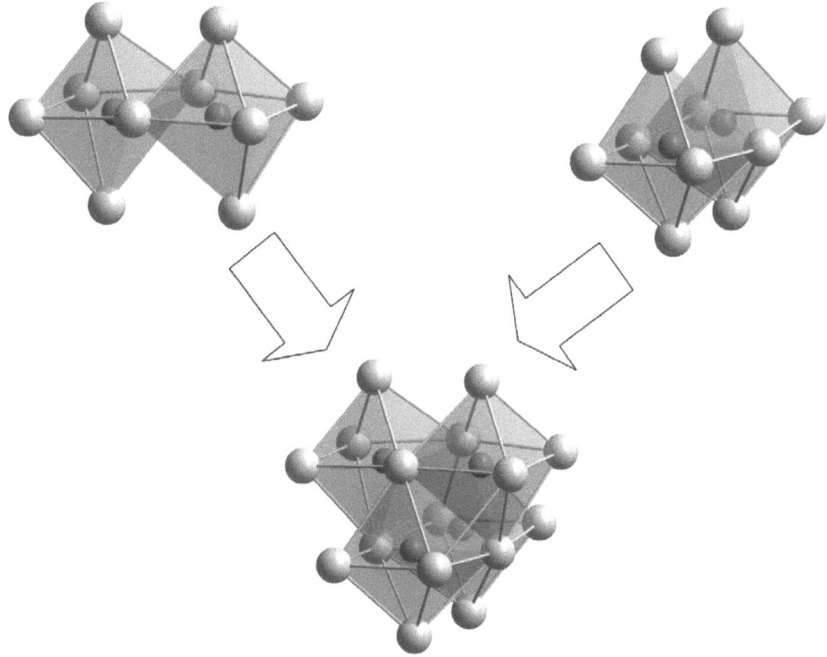

Figure 5.2.: Formation of a $\{Ru_4Ho_{16}\}I_{36}$ tetramer.

On the other hand, these tetrameric clusters can be decribed as a Friauf polyhedron consisting of twelve vertices and with its four hexagonal faces capped by a holmium atom (Fig. 5.3). A Friauf polyhedron is a tetrahedron whose four vertices are truncated.

5. Oligomeric rare-earth clusters

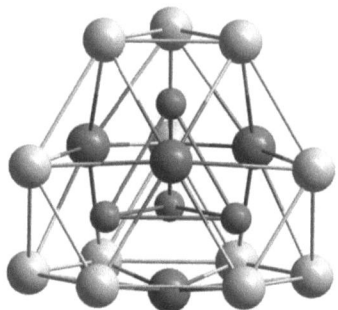

Figure 5.3.: Represenatation of {Ru$_4$Ho$_{16}$} I$_{36}$ as a Friauf polyhedron with six capped hexagonal faces (the capping holmium atoms are marked in dark grey).

The metal framework of {Ru$_4$Ho$_{16}$} I$_{28}$ {Ho$_4$} is illustrated in Fig. 5.4, with the {Ru$_4$Ho$_{16}$} oligomers at the origin position and the Ho$_4$I$_8$ tetrahedra in the body center of the primitive cubic cell. The oligomers are orientated within the cell in a way that the 3-fold axes along the cell diagonals crosses the capped hexagonal faces as well as the truncated Ho-Ho-Ho face of the pseudo-tetrahedron. The {Ho$_4$} units are located on the same axes with its four vertices pointing towards the truncated faces of the Friauf polyhedron.

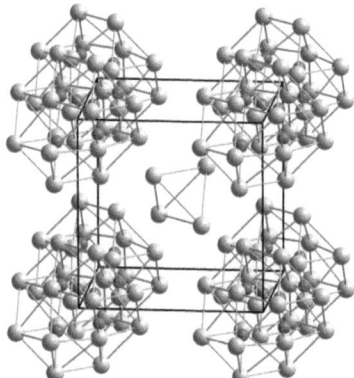

Figure 5.4.: Cubic unit cell of {Ru$_4$Ho$_{16}$} I$_{28}$ {Ho$_4$}. Iodine atoms are omitted for clarity.

The structure consists of three crystallographically inequivalent holmium atoms. The Ho2 atoms define the Friauf polyhedron and are bound to five other Ho atoms while the Ho1 atoms

5. Oligomeric rare-earth clusters

cap its hexagonal faces. Within the oligomer, the Ho1 atoms define another tetrahedron, though its interatomic distances of 437 pm are considerably large (Fig. 5.5). The other Ho-Ho distances are similar to those in other Ho cluster halides and range from 359.38(3) pm to 386.31(3) pm. However, the Ho-Ho distances in the additional empty tedrahedral {Ho$_4$} cluster can be considered as practically nonbonding with a value of 445.52(4) pm. The interatomic distance between the interstial Ru atom and Ho1 is a little larger (280.51 pm) than the corresponding one to Ho2 (269.20(3) pm) because the capping Ho1 atoms lie distinctly outside the Ho2 hexagons.

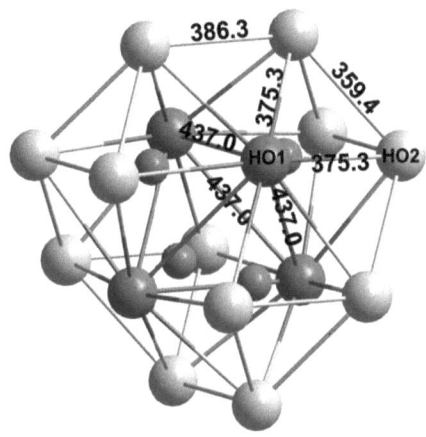

Figure 5.5.: Ho-Ho distances in the oligomeric unit {Ru$_4$Ho$_{16}$}I$_{36}$ (in pm).

The arrangement of the 36 iodine atoms surrounding the metal core is shown in Fig. 5.1. The Ho2 vertices are surrounded by a nearly square array of iodine atoms, although the halogen square lies clearly outside the metal center. In general, the Ho-I bonds can be distinguished in three different modes of iodine bonding (Fig. 5.6). I4 (purple) caps Ho2-Ho2-Ho1 faces and additionally binds exo to an adjacent cluster. As a result, each oligomeric cluster is connected to six other tetramers. The second bonding mode consists of I5 ligands (pink) bridging Ho2-Ho2 edges on the one hand and binding to Ho3 vertices of the empty {Ho$_4$} tetrahedron on the other hand. Hence, the {Ho$_4$} cluster is connected to four oligomers. The third bonding mode involves the I6 atoms (burgundy) only capping the four faces of the {Ho$_4$} tetrahedron.

5. Oligomeric rare-earth clusters

Figure 5.6.: Halogen bridges in $\{Ru_4Ho_{16}\}I_{28}\{Ho_4\}$ with I4 (purple), I5 (pink) and I6 (burgundy).

The Ho-I distances in the oligomeric unit range from 313.44 (3) to 329.51(4) pm for inner bonds whereas they are 348.38(4) pm for the outer Ho1-I4 bonds and therefore being the largest. In the $\{Ho_4\}$ tetrahedron, the Ho3-I6 distances, i.e. comprising the face capping iodine atoms, are 313.65(4) pm and thus practically in the same range as the face capping ones in the macrocluster. The Ho3-I5 distances are considerably short in terms of exo bonds (313.92(4) pm), exhibiting almost the same value as the inner Ho-I bonds.

The empty $\{Ho_4\}$ tetrahedron mentioned above is a sort of extraordinary building unit in the case of holmium halide clusters since it resembles PrI_2-V. As the attempt of synthesizing HoI_2 has failed so far, $\{Ru_4Ho_{16}\}I_{28}\{Ho_4\}$ is the first compound with a substructure corresponding to a possible HoI_2 phase, even though the Ho-Ho distances of 445.52(4) pm are way too large for establishing an "isolated" HoI_2 analogous to PrI_2-V (Fig. 5.7) with Pr-Pr distances of 387.45 pm [68].

5. Oligomeric rare-earth clusters

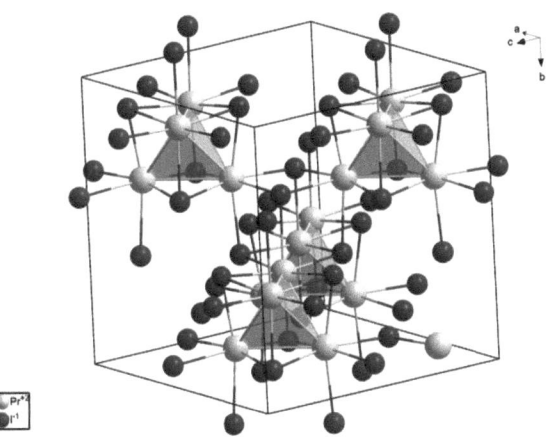

Figure 5.7.: PrI$_2$ – V with structurally the same tetrahedral cluster as in {Ru$_4$Ho$_{16}$} I$_{28}$ {Ho$_4$}.

Synthesis of {Ru$_4$Ho$_{16}$} I$_{28}$ {Ho$_4$}

{Ru$_4$Ho$_{16}$} I$_{28}$ {Ho$_4$} was synthesized using 400 mg HoI$_3$, 91 mg holmium powder and 19 mg ruthenium powder in order to synthesize the monomeric phase Ho {RuHo$_6$} I$_{12}$ according to the reaction 4 HoI$_3$ + 3 Ho + Ru → Ho {RuHo$_6$} I$_{12}$. The reaction mixture was loaded into a tantalum container and exposed to the following temperature program:

$$RT \xrightarrow{50°C/h} 800°C \xrightarrow{15°C/h} 900°C \ (300 \ h) \xrightarrow{2°C/h} 700°C \xrightarrow{50°C/h} RT$$

The product consisted of some black cubic crystals with edge lengths of ca. 0.3 mm and rather large amounts of unreacted HoI$_3$ as well as some HoOI impurities.

The following attempt to yield a pure {Ru$_4$Ho$_{16}$} I$_{28}$ {Ho$_4$} phase by using the appropriate stoichiometric amounts of the reactants was successful: 400 mg HoI$_4$, 138 mg holmium powder and 32 mg ruthenium powder were filled into a niobium container and exposed to the following temperature program:

5. Oligomeric rare-earth clusters

RT $\xrightarrow{50°C/h}$ 800°C $\xrightarrow{15°C/h}$ 950°C (300 h) $\xrightarrow{2°C/h}$ 700°C $\xrightarrow{50°C/h}$ RT

A dark gray powder was yielded with no obvious impurities according to X-ray powder diffraction analysis.

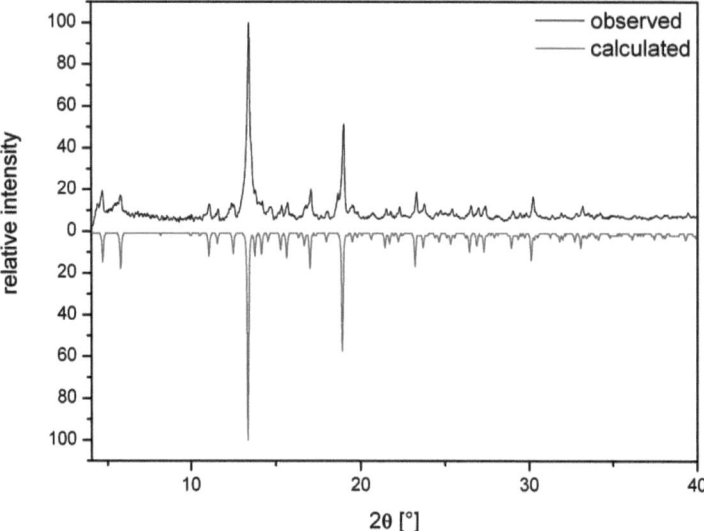

Figure 5.8.: The powder diffraction diagram of the synthesized $\{Ru_4Ho_{16}\}I_{28}\{Ho_4\}$ shows no impurities.

5. Oligomeric rare-earth clusters

Table 5.2.: Crystallographic data of $\{Ru_4Ho_{16}\}I_{28}\{Ho_4\}$

Compound	$\{Ru_4Ho_{16}\}I_{28}\{Ho_4\}$
Cell parameters	a = 1221.14(14) pm
Cell volume	V = 1820.9(4) $\cdot 10^6$ pm^3
Formula units Z	4
Crystal system	cubic
Space group	$P\bar{4}3m$
Instrument	STOE IPDS I
Radiation	MoK$_\alpha$ (λ = 71.07 pm)
Monochromator	graphite
Temperature	293 K
Density	6.617 g/cm^3
F(000)	3000
Absorption correction	numerical
Absorption coefficient	34.103
Number of measured reflections	10978
Number of independent reflections	662
Number of parameters	32
R_{int}	0.1886
Computing structure solution/refinement	SIR-92, SHELX-97
Scattering factors	International Tables, Vol. C
R_1	0.0381 for 436 $F_0 > 4\ \sigma(F_0)$, 0.0678 for all data
wR_2 (for all data)	0.0633
Goodness of fit S	0.752

Table 5.3.: Selected distances in $\{Ru_4Ho_{16}\}I_{28}\{Ho_4\}$

Atom 1	Atom 2	Distance/pm	Atom 1	Atom 2	Distance/pm
Ho3	Ho3	445.52(4)	Ho2	I5	313.99(3)
Ho2	Ho2	386.31(3)	Ho3	I5	313.92(4)
Ho1	Ho2	375.29(3)	Ho3	I6	313.65(4)
Ho2	Ho2	359.38(3)	Ho2	I4	313.44(3)
Ho2	I4	348.38(4)	Ho1	Ru1	280.51(3)
Ru1	Ru1	345.81(3)	Ho2	Ru1	269.20(3)
Ho1	I4	329.51(4)			

5.2. Electronic structure of $\{Ru_4Ho_{16}\}I_{28}\{Ho_4\}$

$\{Ru_4Ho_{16}\}I_{28}\{Ho_4\}$ exhibits 64 electrons or 16 per one interstitial atom. However, the isotypic compound $\{Mn_4Sc_{16}\}Br_{28}\{Sc_4\}$ incorporating the electron poorer Mn consists of 60 (15) electrons which is known to be the optimal electron number according to EHMO calculations of oligomeric $\{Z_4M_{16}\}X_{36}^{x-}$ units as they correlate to a completely filled shell [65]. These results as well as magnetic measurements and inconsistencies in the crystal structure analysis of analogous scandium bromide compounds led to the conclusion that the Sc sites in the empty $\{Sc_4\}$ are slightly under-occupied, leading to the formula $\{Mn_4Sc_{16}\}Br_{28}\{Sc_3\}$ or $Sc_{19}Br_{28}Ru_4$, respectively, in the case of Ru as interstitial. Yet, such an under-occupation could not be found in $\{Ru_4Ho_{16}\}I_{28}\{Ho_4\}$.

In order to get new insights into the electronic structure of $\{Ru_4Ho_{16}\}I_{28}\{Ho_4\}$, DOS and COHP calculations were undertaken. Fig. 5.9 illustrates the projected DOS curves whereas Fig. 5.10 represents the corresponding COHPs. As in the monomeric 7-12 phases, the bonding interactions below the Fermi level are predominantly formed by Ho-Ru interactions while the Ho-Ho bonding interactions are considerably smaller. The rather ionic Ho-I bonding interactions are located at very low energies between -6 and -3 eV, though anti-bonding states are also occupied below the Fermi level. This is consistent with the excessive electrons (64 instead of 60) resulting from the EHMO calculations. A lower electron number would shift the Fermi level to lower energies avoiding the occupancy of anti-bonding Ho-I states. Consequently, a test calculation of the imaginary $\{Tc_4Ho_{16}\}I_{28}\{Ho_4\}$ (60 electrons!) was performed (Fig. 5.11), showing that the Fermi level lies right above the large Ho-Tc bonding interactions, shifting some of the Ho-I anti-bonding states above the Fermi level.

5. Oligomeric rare-earth clusters

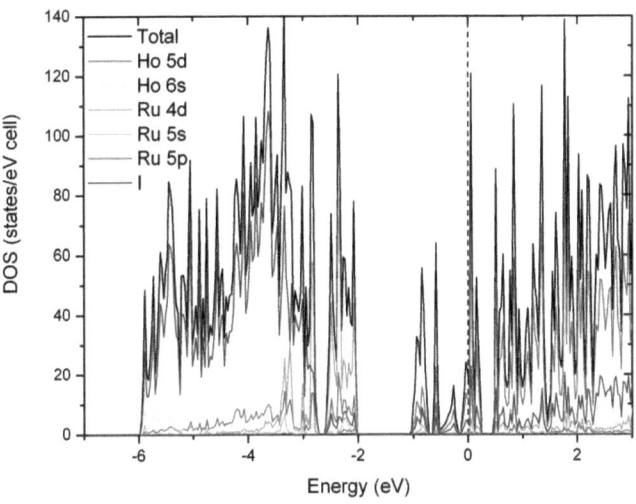

Figure 5.9.: Projected DOS of $\{Ru_4Ho_{16}\}I_{28}\{Ho_4\}$.

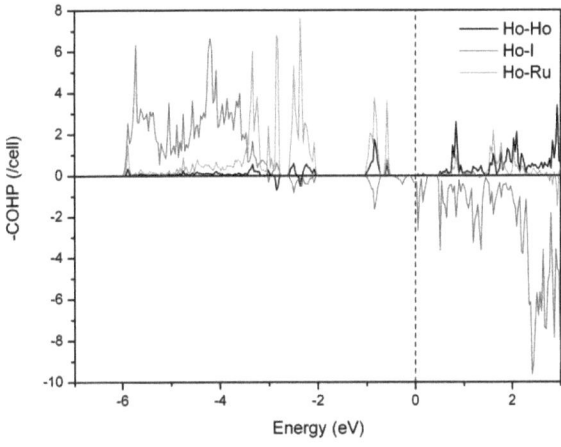

Figure 5.10.: COHP curves of $\{Ru_4Ho_{16}\}I_{28}\{Ho_4\}$.

5. Oligomeric rare-earth clusters

On the basis of the projected DOS of $\{Ru_4Ho_{16}\}I_{28}\{Ho_4\}$, it is possible to assign the orbital characters of the Ho-Ru interactions. Those below -2 eV are mainly built by the interactions of Ho $5d$ and Ru $4d$ orbitals whilst those closer to the Fermi level (-1 to 0.5 eV) are basically formed by Ho $5d$ and Ru $5p$ orbitals. Interestingly, a pseudo band gap separates both contributions.

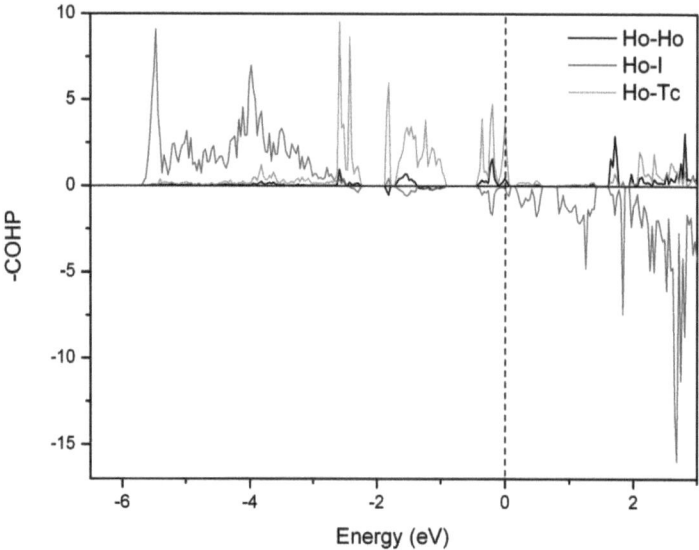

Figure 5.11.: COHP curves of the imaginary $\{Tc_4Ho_{16}\}I_{28}\{Ho_4\}$.

5.3. Crystal structure of $\{(C_2)_2O_2Dy_{14}\}I_{24}$

The example of $\{Ru_4Ho_{16}\}I_{28}\{Ho_4\}$ shows one way of a possible cluster condensation leading to tetramers. Even though the attempt to synthesize the analogous dysprosium compound failed, another tetramer containing dysprosium and main group elements as interstitials could be synthesized: $\{(C_2)_2O_2Dy_{14}\}I_{24}$ crystallizes in the triclinic space group $P\bar{1}$ with the cell parameters a = 972.97(14) pm, b = 1033.03(13) pm, c = 1677.0(2) pm, $\alpha = 101.42(1)°$, $\beta = 92.72(1)°$, $\gamma = 112.75(1)°$ and V = 1509.3(3) $\cdot 10^6$ pm³. Other crystallographic parameters can be found in Tab. 5.4, selected distances are listed in Tab. 5.5. The structure is isotypic with those of

5. Oligomeric rare-earth clusters

$Y_7I_{12}C_2N$ [69] and $Er_7I_{12}C_2N$ [70].

In contrast to $\{Ru_4Ho_{16}\}I_{28}\{Ho_4\}$, the tetramers in $\{(C_2)_2O_2Dy_{14}\}I_{24}$ consist of condensed octahedra and tetrahedra that are arranged in a linear fashion (Fig. 5.12). The octahedra encapsulate C_2 units and show a significant elongation along the C_2 axis whereas the tetrahedra incorporate single oxygen atoms. Two of these tetrahedra are condensed to double tetrahedra via a common edge according to $(Dy_2Dy_{2/2}O)_2 \equiv t_2$. These double tetrahedra are flanked by octahedra via common edges, leading to the formulation $Dy_4Dy_{2/2}C_2 \equiv o_1$ for each octahedron. Thus the characteristic building unit $o_1t_2o_1$ with the composition $[Dy_4Dy_{2/2}C_2][Dy_{2/2}Dy_{2/2}O]_2[Dy_4Dy_{2/2}C_2]$ is generated.

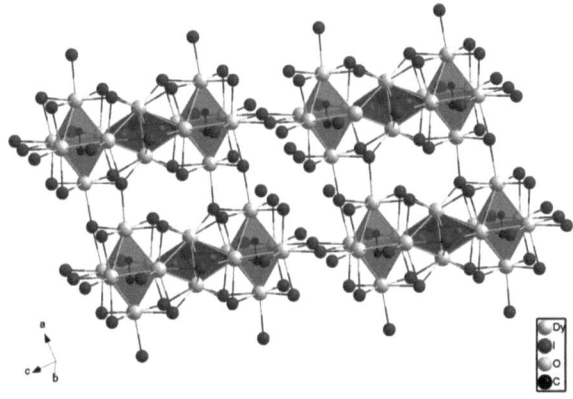

Figure 5.12.: Projection of $\{(C_2)_2O_2Dy_{14}\}I_{24}$ approximately along [010].

The Dy-Dy bond lengths lie in the usual range with significantly shorter distances between dysprosium atoms building up common edges. So the bond lengths of common octahedral/tetrahedral edges are 338.88(5) pm and therefore just marginally shorter than those of the common edges in the double tetrahedra (339.75(4) pm). The other Dy-Dy distances in the tetrahedra range from 364.91(4) to 370.70(4) pm. In contrast, the distances in the octahedra vary much more: Whereas the bond lengths perpendicular to the elongation axis are considerably short (347.63(5) - 351.26(5) pm), the distances along the elongation axis of the

5. Oligomeric rare-earth clusters

octahedron are quite large, as expected (376.08(6) - 389.28(5) pm). The C_2 dumbbell encapsulated in the octahedron has a C-C bond length of 141.06(2) pm which is only slightly larger than a C=C double bond (Fig. 5.13).

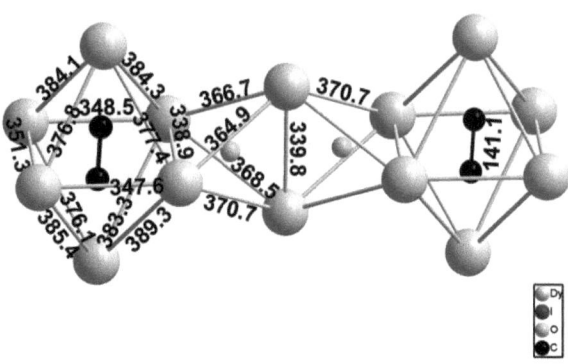

Figure 5.13.: Dy-Dy and C-C distances in $\{(C_2)_2O_2Dy_{14}\}I_{24}$ (in pm). Iodine atoms are omitted for clarity.

The iodine atoms bond to all of the vertices and edges that are not involved in the condensation, with the typical coordination motif M_6X_{12} in case of the octahedron. The Ho-I distances exhibit usual values between 295.44(6) and 333.04(4) pm except for the interaction between the dysprosium atoms involved in the t_2 condensation and the appropriate iodine atoms, exhibiting quite large distances of 369.95(5) and 373.07(6) pm, respectively. The oligomers are connected via I^i and I^a bridges to each other, being established parallel to [100] and [001] according to the formula $\{Dy_4Dy_{2/2}C_2\}_2\{Dy_{2/2}Dy_{2/2}O\}_2 I^{i-i}_{16}I^{i-a}_{8/2}I^{a-i}_{8/2}$. An overview of the connection mode in Fig. 5.14 shows the I^{i-a} contacts in the a/c-plane, but also illustrates that the interactions between these "layers" along [001] only base on van-der-Waals interactions.

5. Oligomeric rare-earth clusters

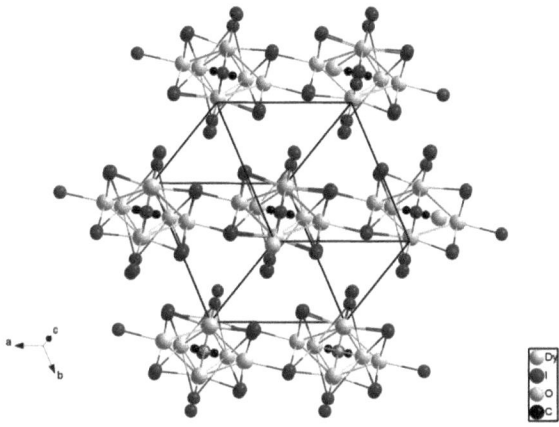

Figure 5.14.: A view at $\{(C_2)_2O_2Dy_{14}\}I_{24}$ along the oligomer's axes.

Synthesis of $\{(C_2)_2O_2Dy_{14}\}I_{24}$

$\{(C_2)_2O_2Dy_{14}\}I_{24}$ was synthesized using 200 mg DyI_3, 84 mg holmium powder, 7 mg Dy_2O_3 and 11 mg graphite powder and NaI as a flux in order to reproduce the chain structure of $\{(C_2)ODy_6\}I_9$. The reaction mixture was loaded in a tantalum container and exposed to the following temperature program:

$$RT \xrightarrow{15°C/h} 1000°C\ (240\ h) \xrightarrow{2°C/h} 600°C \xrightarrow{50°C/h} RT$$

The product consisted of some black, rectangular plates with edge lengths of ca. 0.2 mm which were embedded in the NaI flux. Despite the existence of the previously synthesized compounds $Y_7I_{12}C_2N$ and $Er_7I_{12}C_2N$ which contain nitrogen, the interstitial rather seems to be oxygen in the case of $\{(C_2)_2O_2Dy_{14}\}I_{24}$ since Dy_2O_3 was used as well as an argon atmosphere in the glove box, i.e. no obvious nitrogen source was involved in the reaction process in contrast to the syntheses of $Y_7I_{12}C_2N$ and $Er_7I_{12}C_2N$. Additionally, the structural relationship to the cluster chain $\{(C_2)ODy_6\}I_9$ is remarkable.

Table 5.4.: Crystallographic data of $\{(C_2)_2O_2Dy_{14}\}I_{24}$

Compound	$\{(C_2)_2O_2Dy_{14}\}I_{24}$
Cell parameters	a = 972.97(14) pm
	b = 1033.03(13) pm
	c = 1677.0(2) pm
	α = 101.42(1)°
	β = 92.72(1)°
	γ = 112.75(1)°
Cell volume	V = 1509.3(3) $\cdot 10^6$ pm^3
Formula units Z	2
Crystal system	triclinic
Space group	$P\bar{1}$
Instrument	STOE IPDS I
Radiation	MoK$_\alpha$ (λ = 71.07 pm)
Monochromator	graphite
Temperature	293 K
Density	5.942 g/cm^3
F(000)	2236
Absorption correction	numerical
Absorption coefficient	29.367
Number of measured reflections	29843
Number of independent reflections	8305
Number of parameters	184
R_{int}	0.1886
Computing structure solution/refinement	SIR-92, SHELX-97
Scattering factors	International Tables, Vol. C
R_1	0.0616 for 5659 $F_0 > 4\ \sigma(F_0)$,
	0.0876 for all data
wR$_2$ (for all data)	0.1566
Goodness of fit S	0.941

Table 5.5.: Selected distances in $\{(C_2)_2O_2Dy_{14}\}I_{24}$

Atom 1	Atom 2	Distance/pm	Atom 1	Atom 2	Distance/pm
Dy4	Dy6	389.28(5)	Dy7	I6	333.04(4)
Dy1	Dy4	385.45(4)	Dy1	I9	329.47(4)
Dy2	Dy5	384.29(4)	Dy7	I12	326.41(6)
Dy2	Dy3	384.08(5)	Dy7	I7	325.36(4)
Dy4	Dy5	383.28(5)	Dy6	I10	321.20(6)
Dy2	Dy6	377.44(6)	Dy4	I6	319.09(6)
Dy1	Dy2	376.83(5)	Dy6	I6	318.25(4)
Dy3	Dy4	376.08(6)	Dy5	I12	315.58(4)
Dy7	I10	373.07(6)	Dy2	I12	314.42(4)
Dy6	Dy7	370.70(4)	Dy1	I9	314.00(4)
Dy7	I10	369.95(5)	Dy6	I7	311.50(4)
Dy5	Dy7	368.55(7)	Dy2	I7	311.03(6)
Dy5	Dy7	366.68(6)	Dy1	I11	306.53(6)
Dy6	Dy7	364.91(4)	Dy3	I1	304.16(4)
Dy1	Dy3	351.26(5)	Dy4	I9	303.55(4)
Dy1	Dy6	347.63(5)	Dy5	I2	303.35(6)
Dy7	Dy7	339.75(4)	Dy3	I2	300.76(4)
Dy5	Dy6	338.88(5)	Dy4	I1	295.44(6)
Dy3	I11	335.09(7)	C1	C2	141.06(2)

6. Cluster chains

Further condensation of clusters and the resulting decrease in the X/M ratio leads to the formation of a plethora of structures based on chains, double chains and layers. Most of them consist of metal clusters with encapsulated C, N and C_2 units, but there are also cluster chains intercalated by transition metals, e.g. as found in Pr_4I_5Z [71] or the monoclinic and orthorhombic M_3I_3Z types [72, 73]. In case of dysprosium, a cluster chain of the composition $M_4I_6(C)_2$ has been synthesized in 2007 [24] and could be reproduced in this work. Additionally, the erbium analog was successfully synthesized for the first time. In contrast to dysprosium, a chain structure composed of holmium clusters had been unknown before. Solely a waved double layer, $[Ho_9C_4O]I_8$, consisting of holmium octahedra and tetrahedra comprising C and O atoms, respectively, has been synthesized in 1993 (Fig. 6.1) [74].

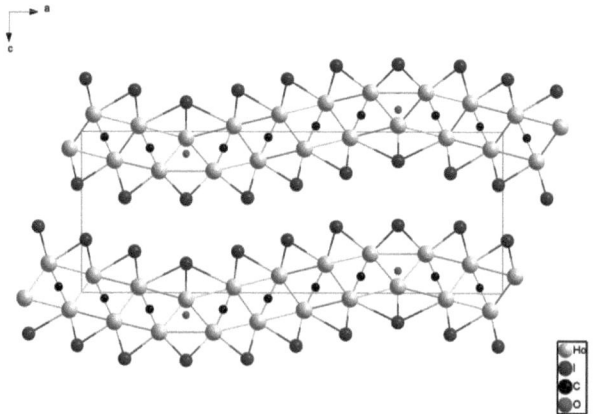

Figure 6.1.: Projection of a part of the crystal structure of $[Ho_9C_4O]I_8$ onto [010].

6. Cluster chains

6.1. Crystal structure of $\{(C_2)ODy_6\}I_9$

As seen before, the smallest possible unit consisting of $\{(C_2)Dy_6\}$ octahedra and $\{O_2Dy_6\}$ double tetrahedra, i.e. $o_1t_2o_1$, is realized in the oligomeric $\{Dy_{14}(C_2)_2O_2\}I_{24}$. Even larger fragments $(o_1t_2o_1)_n$ are imaginable by adding n oligomers of o_2t_2. Indeed, with $\{(C_2)ODy_6\}I_9$ the other extreme $\frac{1}{\infty}[o_2t_2]$ can be formed. $\{(C_2)ODy_6\}I_9$ crytallizes in the hexagonal space group $P6/m$ with the cell parameters a = 2024.18(8) pm, c = 1299.21(4) pm and V = 4610.1(3) ·10^6 pm^3. Other crystallographic parameters can be found in Tab. 6.1, selected distances are listed in Tab. 6.2.

The principal motif of the chain is shown in Fig. 6.2. The only obvious difference from the oligomer $\{Dy_{14}(C_2)_2O_2\}I_{24}$ is an additional common edge between the condensed octahedra and hence the associated iodine ligands. Likewise, the interatomic distances are very similar to those in the oligomer, only the bond length of the additional common edge in the chain clusters is distinctly shorter (319.32(15) pm) due to the condensation (Fig. 6.2). The C-C distances (144(2) pm) are a little larger than those in the oligomer, but still range between a single and double bond.

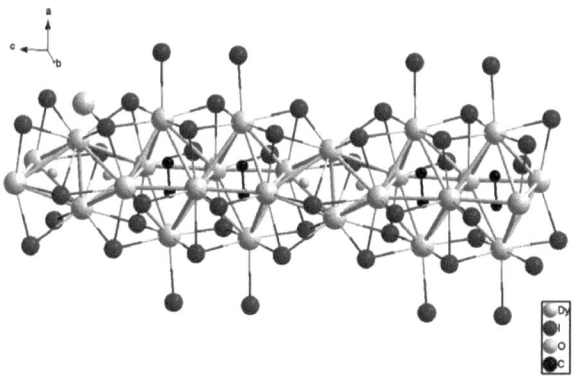

Figure 6.2.: Cluster chain in $\{(C_2)ODy_6\}I_9$.

6. Cluster chains

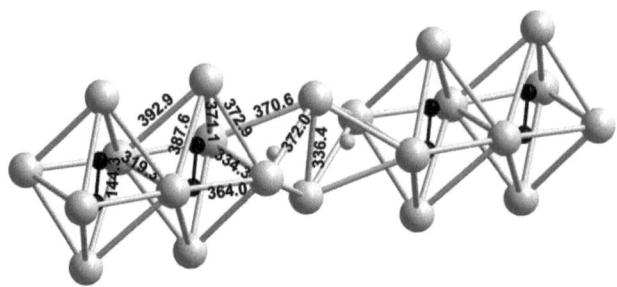

Figure 6.3.: Dy-Dy and C-C distances in $\{(C_2)ODy_6\}I_9$.

The chains are connected to each other by I^{a-i} (o-o) contacts, leading to the formation of hexagonal tubes along the c-axis with a diameter of ca. 1320 pm, incorporating another chain composed of the same motif as the former ones. The existence of three energetically equivalent orientations of the isolated chain causes a corresponding disorder in the crystal structure (Fig. 6.4). Even if the general structure of this chain consists of the $\frac{1}{\infty}[o_2t_2]$ arrangement, there are slight deviations concerning some interatomic distances and the C-C alignments (Fig. 6.5). The latter are pointed towards the faces of the octahedra, in contrast to the corresponding ones in the non-disordered chains, in which they point towards the vertices. As a consequence, the C-C distance (153(5) pm) is much large as the repulsion between the Dy and C atoms decreases when a diagonal orientation is established. Besides, it is notable that the carbon positions are not determined accurately in an environment of heavy atoms. Another remarkable difference is that the octahedral vertices in the disordered chain are not really bonded to iodine atoms, i.e. they do not possess terminal ligands. Whereas these distances exhibit a value of 331 pm in the non-disordered chain, the corresponding Dy-I distance, constructed by the interaction of a vertex Dy atom and an inner iodide ligand of a non-disordered chain, is more than 370 pm and thus just a very weak interaction.

6. Cluster chains

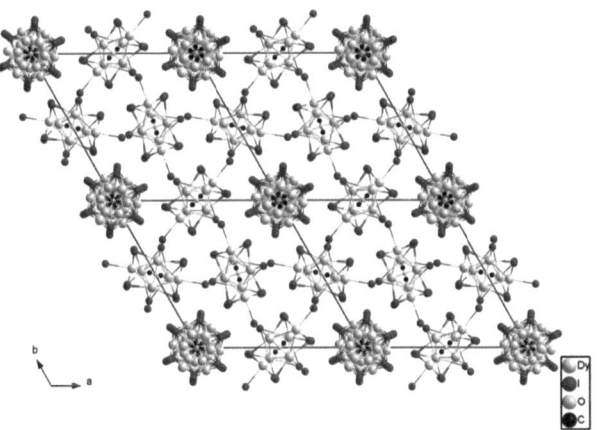

Figure 6.4.: View at a projection of the unit cell of $\{(C_2)ODy_6\}I_9$ along [001].

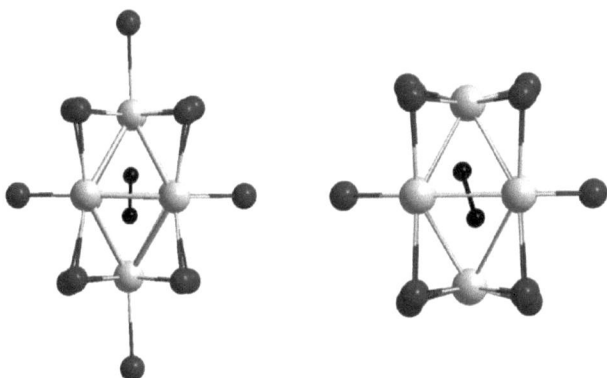

Figure 6.5.: Comparison of the Dy octahedra in the non-disordered and disordered chains.

$\{(C_2)ODy_6\}I_9$ is not the first cluster chain composed of metal octahedra and tetrahedra. In 1994 Mattausch et al. reported about the tetragonal $\alpha - Gd_4I_6CN$ and hexagonal $\beta - Gd_4I_6CN$ structure types, both comprising the polymeric $\frac{1}{\infty}[o_1t_2]$ chains, i.e. not exhibiting double octahedra [75]. And they do not contain oxygen atoms as interstitials in the tetrahedral voids, but nitrogen atoms instead. Like $\{(C_2)ODy_6\}I_9$, the hexagonal $\beta - Gd_4I_6CN$ phase exhibits a

6. Cluster chains

disordered chain located in a hexagonal tube. Furthermore, they reported about the analogous $Y_6I_9(C_2)N$, but solved the structure in space group $P6$.

Synthesis of $\{(C_2)ODy_6\}I_9$

$\{(C_2)ODy_6\}I_9$ was synthesized using 150 mg DyI_3, 63 mg dysprosium powder, 8 mg graphite powder and 10 mg iron powder, so it was actually intended to synthesize a dysprosium halide cluster with alternating carbon and iron atoms as interstitials. The reaction mixture was loaded into a tantalum container and exposed to the following temperature program:

$$RT \xrightarrow{50°C/h} 800°C \xrightarrow{15°C/h} 1000°C \text{ (200 h)} \xrightarrow{2°C/h} 680°C \xrightarrow{50°C/h} RT$$

The product mixture mostly contained black needles that easily split upon applying mechanical pressure. Even though no oxygen source was used, oxygen impurities are abundant, either as atmospheric impurities in the glove box, as contaminated reactants or reaction containers. The unintended synthesis of $\{(C_2)ODy_6\}I_9$ resulted in attempts to reproduce this phase. However, those attempts led to the synthesis of $\{(C_2)_2O_2Dy_{14}\}I_{24}$ (section 5.3).

6. Cluster chains

Table 6.1.: Crystallographic data of $\{(C_2)ODy_6\}I_9$

Compound	$\{(C_2)ODy_6\}I_9$
Cell parameters	a = 2024.18(8) pm
	c = 1299.21(4) pm
Cell volume	V = 4610.1(3) $\cdot 10^6$ pm^3
Formula units Z	8
Crystal system	hexagonal
Space group	$P6/m$
Instrument	STOE IPDS I
Radiation	MoK$_\alpha$ (λ = 71.07 pm)
Monochromator	graphite
Temperature	293 K
Density	6.158 g/cm^3
F(000)	7064
Absorption correction	numerical
Absorption coefficient	31.236
Number of measured reflections	62811
Number of independent reflections	3526
Number of parameters	149
R_{int}	0.1070
Computing structure solution/refinement	SIR-92, SHELX-97
Scattering factors	International Tables, Vol. C
R_1	0.0344 for 2594 $F_0 > 4\ \sigma(F_0)$,
	0.0550 for all data
wR_2 (for all data)	0.0830
Goodness of fit S	0.979

Table 6.2.: Selected distances in $\{(C_2)ODy_6\}I_9$

Atom 1	Atom 2	Distance/pm	Atom 1	Atom 2	Distance/pm
Dy1	Dy3	392.92(9)	Dy11	I21	325.60(12)
Dy1	Dy3	387.63(8)	Dy11	I18	323.36(11)
Dy1	Dy2	372.90(8)	Dy3	I14	322.26(14)
Dy11	Dy2	372.01(8)	Dy3	Dy3	319.32(15)
Dy2	Dy1	371.06(8)	Dy2	I21	316.45(11)
Dy11	Dy2	370.55(8)	Dy3	I15	314.17(13)
Dy3	Dy2	364.04(5)	Dy2	I18	312.82(10)
Dy31	I34	342.4(3)	Dy1	I14	305.95(10)
Dy11	Dy11	336.43(18)	Dy1	I21	305.8(1)
Dy2	Dy2	334.33(11)	Dy1	I18	305.42(10)
Dy1	I10	331.01(9)	Dy2	I10	304.66(9)
Dy3	I10	327.6(1)	Dy1	I15	302.92(9)
Dy2	I29	327.43(8)	C2	C2	153(5)
Dy2	I29	327.43(8)	C1	C1	144(2)

6.2. Crystal structure of $\{IrHo_3\}I_3$

The trend of favorably encapsulating main group elements in dysprosium clusters is followed in the $\{(C_2)ODy_6\}I_9$ compound above. Attempts to synthesize an analogous holmium cluster resulted in the formation of $[Ho_9C_4O]I_8$ (Fig. 6.1). On the other hand, an attempt to insert a transition metal into the holmium cluster yielded compound $\{IrHo_3\}I_3$ which does not contain a single chain, but a double chain, further condensed side-by-side as it is similarly realized in other systems such as $\{C_2Sc_7\}Cl_{10}$ [76] and $\{C_2Y_6\}I_7$ [77] for instance. Also the isotypic compounds $\{RuLa_3\}I_3$, $\{RuPr_3\}I_3$, $\{RuGd_3\}I_3$, $\{RuY_3\}I_3$, $\{RuEr_3\}I_3$, $\{IrGd_3\}I_3$ and $\{IrY_3\}I_3$ have been synthesized by Corbett and co-workers [72].

$\{IrHo_3\}I_3$ crystallizes in the monoclinic space group $P2_1/m$ with the cell parameters a = 872.0(4) pm, b = 423.83(14) pm, c = 1211.2(6) pm, $\beta = 94.99(4)°$ and V = 445.9(3) $\cdot 10^6$ pm^3. Other crystallographic parameters can be found in Tab. 6.3, selected distances are listed in Tab. 6.4. A section of the double chain is illustrated in Fig. 6.6. The endohedral Ir atoms are surrounded by seven holmium atoms forming monocapped trigonal prisms, sharing common

6. Cluster chains

rectangular faces and thus construct chains according to the formulation $\{\text{IrHo}_{6/3}\text{Ho}_{1/1}\}\text{I}_3$.

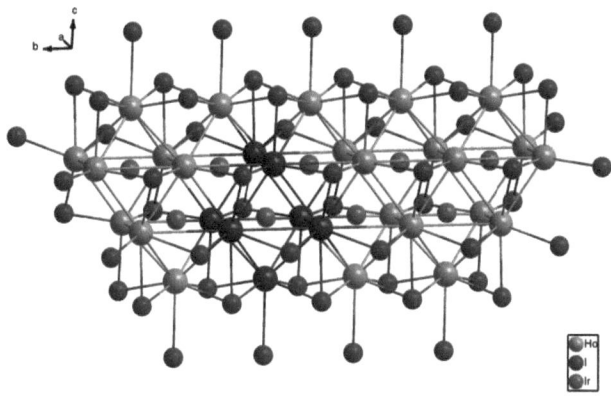

Figure 6.6.: Side view of the double chain in $\{\text{IrHo}_3\}\text{I}_3$. The trigonal-prismatic building unit is highlighted.

The trigonal prisms are heavily distorted which is also evident in the pretty diverse Ho-Ir distances, with an average of 287 pm:

Ir Ho2: 275.9(10) pm

Ir Ho1 (x2): 277.3(5) pm

Ir Ho3 (x2): 277.2(6) pm

Ir Ho1: 291.1(10) pm

Ir Ho3: 336.0(9) pm

6. Cluster chains

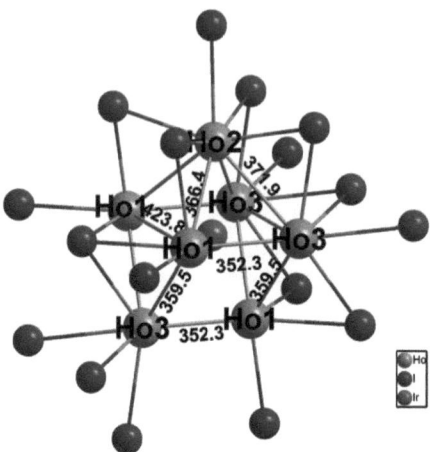

Figure 6.7.: Monocapped prism in {IrHo3}I3

Due to these large differences, the clusters can also be described as edge-sharing square (rectangular) pyramidal IrHo$_5$ units, only involving the five shortest, almost equal Ho-Ir distances, with the Ir interstitials exhibiting a slight out-of-plane arrangement, i.e. below the basal plane.

In contrast to that, the Ho-Ho distances rather suggest a description as a monocapped trigonal prism. Whereas the rectangular base exhibits quite unequal distances of 423.83(14) and 352.3(9) pm, respectively, the Ho-Ho bonds constructing the rectangular faces as well as those forming the capping are very similar (352.3(9) - 371.9(9) pm) (Fig. 6.8).

In general, the seven-coordinated monocapped trigonal prismatic arrangement in {IrHo3}I3 is the consequence of an extreme distortion of edge-sharing IrHo$_6$ octahedra as they exist in the {RuPr3}I3 relative [72].

The metal chains are surrounded by iodine atoms, connecting neighboring chains among each other. The iodine atoms can be divided into three crystallographically different groups (Fig. 6.9): The I1 atoms bridge edges of the basal plane of one chain to vertices of the basal plane of another neighboring chain. The I3 atoms also interconnect chains via an I^{i-a} mode, but

6. Cluster chains

bonding to an edge involving the capping Ho atom and to the capping Ho vertex itself from another chain. The I2 atoms, however, do not act as chain interconnecting ligands, but merely bridge one of the edges between the capping Ho atom and the basal plane of the same cluster chain. The Ho-I distances exhibit expected values between 313.3(8)- 347.51(94) pm for inner bonds and 327.5(9)-345.2(13) pm for terminal bonds. Even though they range in the expected range, the large Ho-I distances of the inner Ho3-I1 bridges are significant, but they go along with the relatively large Ho-Ho distances (423.83(14) pm) at those edges.

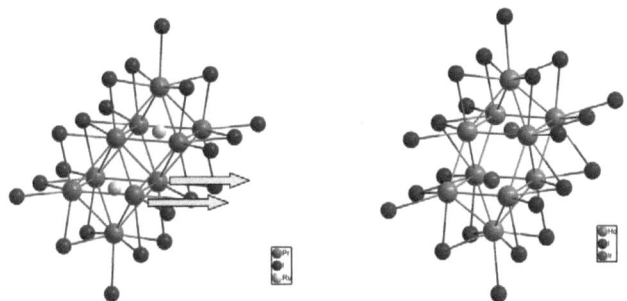

Figure 6.8.: Comparison of double octahedra/prisms in $\{RuPr_3\}I_3$ (left) and $\{IrHo_3\}I_3$ (right). The arrows indicate the distortion leading to the prisms in $\{IrHo_3\}I_3$.

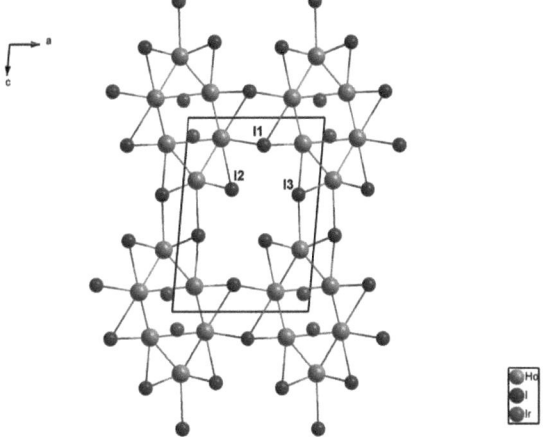

Figure 6.9.: Structure of $\{IrHo_3\}I_3$ viewed along the chains parallel to [010].

The Ir-Ir distances are 313.0(9) pm, so they just exhibit only weak interactions which is also apparent in the correspondent COHP diagram (Fig. 6.11).

Synthesis of $\{\text{IrHo}_3\}\text{I}_3$

$\{\text{IrHo}_3\}\text{I}_3$ was synthesized using 300 mg HoI$_3$, 68 mg holmium powder, 27 mg iridium powder according to the reaction 4 HoI$_3$ + 3 Ho + Ir → Ho$\{\text{IrHo}_6\}\text{I}_{12}$, i.e. it was actually aimed to yield a monomeric cluster. The reaction mixture was loaded into a niobium container and exposed to the following temperature program:

$$\text{RT} \xrightarrow{50°\text{C/h}} 800°\text{C} \xrightarrow{15°\text{C/h}} 950°\text{C } (300 \text{ h}) \xrightarrow{2°\text{C/h}} 700°\text{C} \xrightarrow{50°\text{C/h}} \text{RT}$$

The product mixture contained tiny, black needles as well as traces of HoOI and unreacted HoI$_3$. Even though some crystals were chosen and checked concerning their quality, it was not possible to select a single crystal of a better quality. Further attempts to gain better crystals as well as the pure $\{\text{IrHo}_3\}\text{I}_3$ phase using stoichiometric amounts failed so far.

Table 6.3.: Crystallographic data of $\{IrHo_3\}I_3$

Compound	$\{IrHo_3\}I_3$
Cell parameters	a = 872.0(4) pm
	b = 423.83(14) pm
	c = 1211.2(6) pm
	β = 94.99(4)°
Cell volume	V = 445.9(3) $\cdot 10^6$ pm^3
Formula units Z	2
Crystal system	monoclinic
Space group	$P2_1/m$
Instrument	STOE IPDS I
Radiation	MoK$_\alpha$ (λ = 71.07 pm)
Monochromator	graphite
Temperature	293 K
Density	7.952 g/cm^3
F(000)	874
Absorption correction	numerical
Absorption coefficient	51.484
Number of measured reflections	5058
Number of independent reflections	1181
Number of parameters	43
R_{int}	0.2610
Computing structure solution/refinement	SIR-92, SHELX-97
Scattering factors	International Tables, Vol. C
R_1	0.1898 for 706 $F_0 > 4\ \sigma(F_0)$,
	0.2550 for all data
wR$_2$ (for all data)	0.4438
Goodness of fit S	1.126

6. Cluster chains

Table 6.4.: Selected distances in $\{IrHo_3\}I_3$

Atom 1	Atom 2	Distance/pm	Atom 1	Atom 2	Distance/pm
Ho2	Ho3	371.9(9)	Ho3	I2	326.2(13)
Ho1	Ho2	366.3(8)	Ho2	I3	319.8(9)
Ho1	Ho3	359.5(8)	Ho1	I3	318.1(12)
Ho1	Ho3	352.3(9)	Ho2	I2	313.3(8)
Ho3	I1	347.5(10)	Ir1	Ir1	313.0(9)
Ho2	I3	345.2(13)	Ho1	Ir1	291.1(10)
Ho3	I1	342.7(13)	Ho1	Ir1	277.3(5)
Ho3	Ir1	336.0(9)	Ho3	Ir1	277.2(6)
Ho1	I1	327.5(9)	Ho2	Ir1	275.9(10)

6.3. Electronic structure of $\{IrHo_3\}I_3$

The calculated DOS and COHP curves of $\{IrHo_3\}I_3$ are shown in Figs. 6.10 and 6.11. The trend that Ho-Ho interactions poorly contribute to the cluster formation is obvious again. The bonding Ho-Ho interactions are extremely small below the Fermi level and are distinctly topped by large Ho-Ir bonding contributions. Besides the Ho-I bonding interactions at energies between -6.5 and -1.5 eV and some minor anti-bonding contributions near the Fermi level, the Ir-Ir interactions complete the COHP analysis. The latter slightly resembles a distorted version of a classical filled metal-metal band with bonding interactions at the bottom and anti-bonding ones at the top. In general, the bonding Ir-Ir interactions predominate the states below the Fermi level in comparison with the anti-bonding Ir-Ir interactions.

A further analysis of the orbital character of the Ho-Ir interactions reveals that Ho $5d$ and Ir $5d$ interactions are responsible for the bonding Ho-Ir contributions between -7 and -2 eV. Again, a tiny pseudo band gap at -1.75 eV separates them from the bonding Ho-Ir at the Fermi level, resulting from Ho $5d$ – Ir $6p$ interactions.

6. Cluster chains

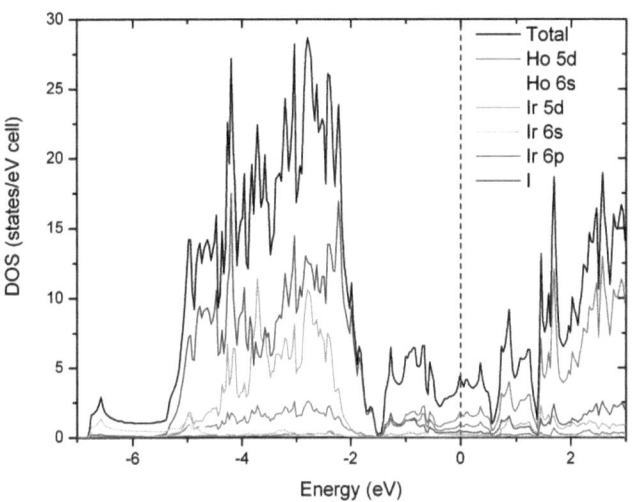

Figure 6.10.: Projected DOS of {IrHo$_3$}I$_3$.

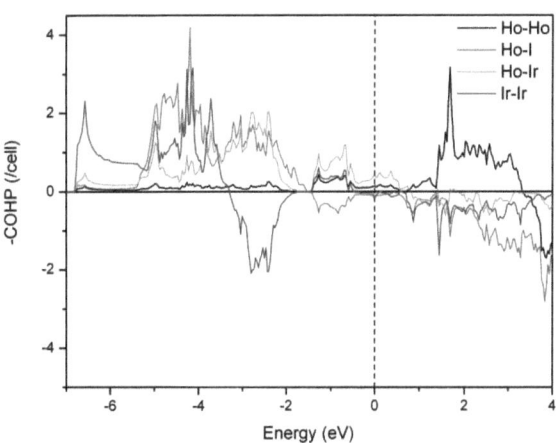

Figure 6.11.: COHP curves of {IrHo$_3$}I$_3$.

6.4. Crystal structure of $\{(C_2)Er_4\}I_6$

In contrast to its neighbor dysprosium, holmium rather incorporates transition metals. Going further to erbium, one can find both, transition metals and main group elements, as interstitials, but all the synthesized erbium clusters have been monomeric, dimeric or oligomeric so far. The only exceptions are R_4I_5 and R_7I_{10} which have been discussed as single (R_4I_5) and double (R_4I_5) strings of edge-sharing erbium octahedra [78]. With $\{(C_2)Er_4\}I_6$, the first interstitially stabilized cluster chain, involving erbium could be synthesized. The compound $\{(C_2)Er_4\}I_6$ crystallizes in the orthorhombic space group $Pnnm$ with the cell parameters a = 1354.71(14) pm, b = 1421.14(17) pm, c = 854.87(8) pm and V = 1645.8(3) $\cdot 10^6$ pm^3. Other crystallographic parameters can be found in Tab. 6.5, selected distances are listed in Tab. 6.6.

The structure derives from the NaMo$_4$O$_6$ structure type comprising empty Mo$_6$ octahedra [79]. Fig. 6.12 shows a section of the erbium chain which consists of distorted $\{(C_2)Er_6\}$ octahedra alternately elongated and compressed according to the orientations of the encapsulated C$_2$ units. I.e., the C$_2$ units are aligned perpendicular to the chain axis in the former and parallel to it in the latter.

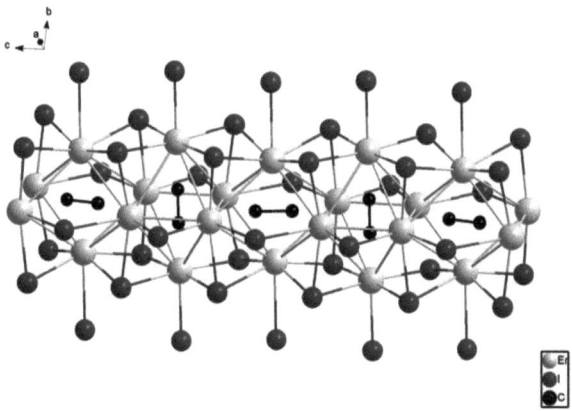

Figure 6.12.: Section of the cluster chain in $\{(C_2)Er_4\}I_6$

The octahedra are connected via common trans edges and exhibit Er-Er distances between

330.50(12) and 379.77(14) pm, taking into account that the compressed octahedron can actually not be described as an octahedron or even distorted octahedron due to extremely large Er-Er distances (497.02(20) pm) parallel to the C_2 units (Fig. 6.13). The C-C distances are 149(4) and 161(9) pm, but these values cannot be considered as precise due to the large standard deviation which is caused by the lack of better crystallographic data and a disorder of the C_2 dumbbell in the compressed octahedron as well as the environment of heavy atoms. There is also a disorder of the Er2 atom, but as the site occupation factor of Er2A is 0.84, the Er2B atom is neglected in the discussion.

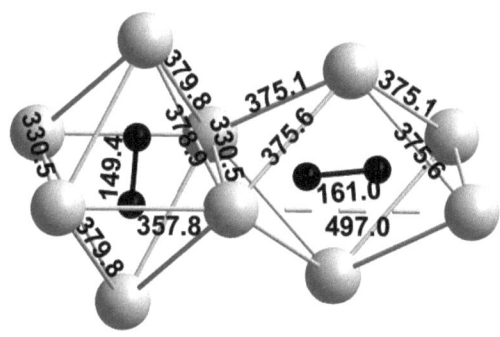

Figure 6.13.: Er-Er and C-C distances in $\{(C_2)Er_4\}I_6$ (in pm).

The metal chains run along [001] and are connected via I^{i-a} and I^{a-i} contacts (I2 and I4) according to the formulation $\{(C_2)Er_4\}I^i_{2/1}I^{i-a}_{4/2}I^{a-i}_{4/2}$ (Fig. 6.14). The I1 and I3 atoms act as bridging inner ligands, leading to a M_6X_{12} analogous halogen surrounding of the metal core. The Ho-I distances exhibit expected values between 304.83(15) and 346.83(42) pm.

6. Cluster chains

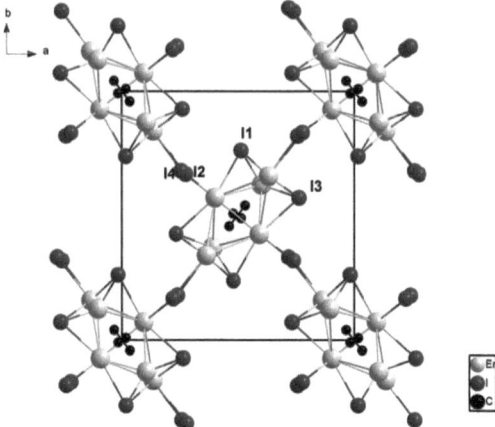

Figure 6.14.: Projection of the crystal structure of $\{(C_2)Er_4\}I_6$ along [001] with emphasis on the Er-I interchain connections.

Synthesis of $\{(C_2)Er_4\}I_6$

Some black, needle shaped crystals of $\{(C_2)Er_4\}I_6$ were found in the same reaction container used for synthesizing $\{(C_2)_2Er_{10}\}I_{18}$ (section 4.2). The yield was low though, but neither a bigger yield nor a pure $\{(C_2)Er_4\}I_6$ phase could be achieved yet.

6. Cluster chains

Table 6.5.: Crystallographic data of $\{(C_2)Er_4\}I_6$

Compound	$\{(C_2)Er_4\}I_6$
Cell parameters	a = 1354.71(14) pm
	b = 1421.14(17) pm
	c = 854.87(8) pm
Cell volume	V = 1645.8(3) $\cdot 10^6$ pm^3
Formula units Z	4
Crystal system	orthorhombic
Space group	$P2_1/m$
Instrument	STOE IPDS I
Radiation	MoK$_\alpha$ (λ = 71.07 pm)
Monochromator	graphite
Temperature	293 K
Density	5.870 g/cm^3
F(000)	2408
Absorption correction	numerical
Absorption coefficient	31.401
Number of measured reflections	16581
Number of independent reflections	2461
Number of parameters	67
R_{int}	0.0867
Computing structure solution/refinement	SIR-92, SHELX-97
Scattering factors	International Tables, Vol. C
R_1	0.0596 for 1553 $F_0 > 4\ \sigma(F_0)$,
	0.0869 for all data
wR_2 (for all data)	0.1637
Goodness of fit S	1.048

Table 6.6.: Selected distances in $\{(C_2)Er_4\}I_6$

Atom 1	Atom 2	Distance/pm	Atom 1	Atom 2	Distance/pm
Er1	Er3	379.77(14)	Er2A	I3	311.01(25)
Er1	Er3	378.93(14)	Er3	I3	310.71(14)
Er2A	Er3	375.62(34)	Er3	I1	310.21(14)
Er2A	Er3	375.13(24)	Er3	I4	307.01(18)
Er3	Er3	357.85(20)	Er1	I1	305.38(15)
Er2A	I4	346.83(42)	Er1	I3	304.83(15)
Er3	Er3	330.50(12)	C2	C2	161(9)
Er3	I2	330.01(18)	C2	C2	152(8)
Er1	I2	314.60(22)	C1	C1	149(4)
Er2A	I1	311.33(31)			

6.5. Electronic structure of $\{(C_2)M_4\}I_6$ type compounds

The electronic structure of $\{(C_2)M_4\}I_6$ type compounds were calculated using the data from $\{(C_2)Dy_4\}I_6$ as the erbium analog is a little disordered.

Figs. 6.15 and 6.16 show the DOS and COHP curves. The latter illustrates the distinct Dy-C bonding interactions below and at the Fermi level. They clearly outweigh the bonding Ho-Ho interactions that are quite small below the Fermi level and are by far not completely filled. The Ho-I bonding interactions are located at lower energies between -5 and -1 eV followed by some anti-bonding Ho-I contributions at the Fermi level. Accoding to the projected DOS, the Ho-C interactions are predominantly formed by Ho $5d$ and C $2p$ orbitals, though some Ho $6s$ contributions are evident in the low energy range around -5 eV.

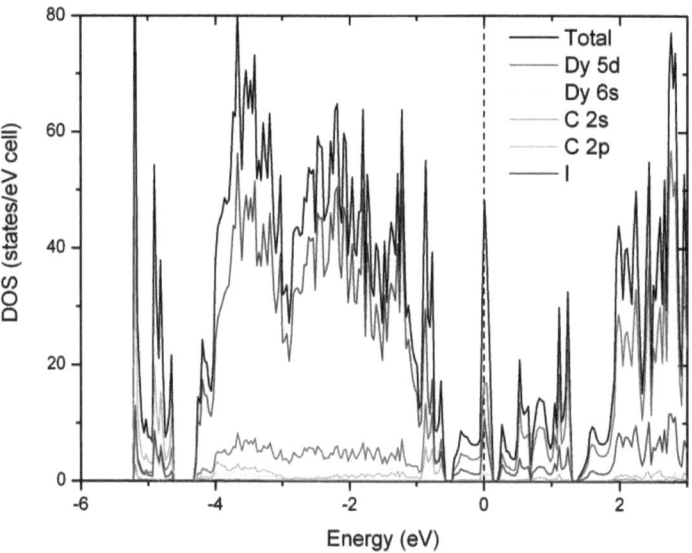

Figure 6.15.: Projected DOS of $\{C_2Dy_4\}I_6$.

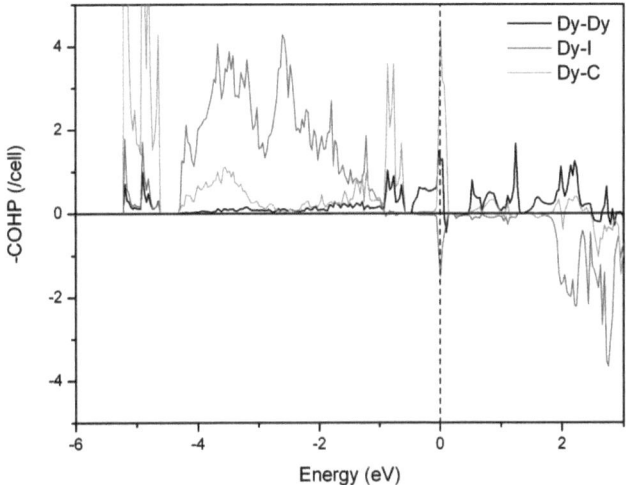

Figure 6.16.: COHP curves of $\{C_2Dy_4\}I_6$.

7. Summary and prospects

The challenge of synthesizing rare-earth halide cluster complexes of the elements dysprosium and holmium led to a plethora of new cluster compounds (all listed below), especially in the field of the monomeric 7-12 type structures. Holmium clusters stabilized by an interstitial transition metal could be obtained for the first time. In this regard, $3d$ as well as $4d$ and $5d$ transition metals can appear as interstitial atoms as it is realized in the monomeric $\text{Ho}\{\text{ZHo}_6\}\text{I}_{12}$ phases with Z = Fe, Co, Ni, Ir and Pt and in the tetrameric $\{\text{Ru}_4\text{Ho}_{16}\}\text{I}_{28}\{\text{Ho}_4\}$.

Commonly, a coordination number of six is preferred for the endohedral transition metal, but bigger metal atoms such as iridium are known to exhibit even larger coordination numbers of seven or eight. This is illustrated by the compound $\{\text{IrHo}_3\}\text{I}_3$ where the iridium atom is coordinated by seven holmium atoms in the form of a monocapped trigonal prism instead of a usually observed octahedron. Therefore a more detailed analysis of inserting heavier transition metals such as osmium and platinum could give further insights into the structural influence of the interstitial on the coordination geometry, but first attempts have not been successful yet. Also, the spectrum of inserted $4d$ elements is very small as it only succeeded to encapsulate ruthenium atoms, so the holmium chemistry offers still prospects of new (unpredictable) halide clusters.

In contrast to the very diverse interstitials in holmium clusters, the corresponding results of dysprosium clusters reveal a pretty small variety since predominantly carbon and oxygen atoms act as interstitials. Numerous attempts to insert transition metals into dysprosium halide clusters failed. Accordingly, main group interstitials dominate the chemistry of dysprosium clusters, building up monomers as in $\text{Dy}\{(\text{C}_2)\text{Dy}_6\}\text{I}_{12}$, dimers as in $\{\text{Dy}_{10}(\text{C}_2)_2\}\text{Br}_{18}$, oligomers as in $\{(\text{C}_2)_2\text{O}_2\text{Dy}_{14}\}\text{I}_{24}$ or chains as in $\{(\text{C}_2)\text{ODy}_6\}\text{I}_9$ or $\{(\text{C}_2)\text{Dy}_4\}\text{I}_6$. And the structural field

7. Summary and prospects

is by far not exploited as crystals presumably of the composition $\{(C_2)_4O_2Dy_{16}\}I_{14}$ could be obtained, consisting of waved layers like in $\{(C_2)_4O_2Y_{18}\}I_{16}$ [80]. But due to poor crystallographic data, the exact composition could not be determined yet.

Even though it was not successful to obtain transition metals as interstitials for dysprosium halide clusters – despite its almost same size as holmium or yttrium atoms whose clusters encapsulate transition metals very easily – a mixed dysprosium/yttrium cluster, namely $Dy\{CoDy_{4.53}Y_{1.47}\}I_{12}$, could be obtained. The fact that the additional yttrium atoms are inserted on the cluster positions backs the idea that something must be different in mere $\{Dy_6\}$ clusters, maybe resulting from electronic reasons. However, a further study of the minimum ratio of Y/Dy that needs to be present in this cluster could provide a better insight into this "dilemma". The reason for the pretty different dysprosium chemistry could be caused by the electronic situation. But as it is only one f-electron that causes all those differences, it was not possible to pursue deeper investigations because the f-electrons had to be treated as core electrons in all the performed calculations.

To extend the knowledge of bonding in rare-earth halide clusters, band structure calculations were applied. On the basis of DOS and COHP analyses, it could be shown that the driving force of forming such cluster halides is not the M-M interactions between rare-earth metals, but the bonding interactions between the interstitial and the rare-earth metal surrounding. This has led to the concept of anti-Werner complexes introduced by Meyer [23]. A similar phenomenon has been observed for rare-earth tellurides [81], although these compounds exhibit slightly more bonding M-M interactions than the rare-earth halide clusters, but this can be caused by the more polar M-X bonds in the case of halides leading to an even greater electron deficiency at the rare-earth metal atoms.

There is still a lot to discover in the chemistry of dysprosium and holmium halide clusters and a closer look at the neighboring erbium is also worth a further investigation. First steps in synthesizing erbium halide clusters yielded the dimer $\{(C_2)_2Er_{10}\}I_{18}$ and the cluster chain $\{(C_2)Er_4\}I_6$. Also, a powder phase suggesting the formation of $Er\{CoEr_6\}I_{12}$ could be obtained which illustrates that a wide variety of interstitials may be found in erbium compounds.

7. Summary and prospects

Summary of synthesized compounds

Ho{ZHo$_6$}I$_{12}$ with Z = Fe, Ni, Pt

Fe:
a = 1529.73(17) pm
c = 1062.52(16) pm
V = 2153.3(5) ·10^6 pm^3
Z = 3
trigonal $R\bar{3}$

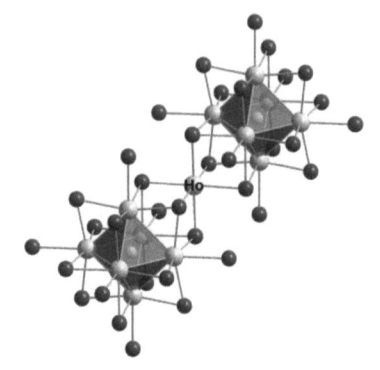

Ni:
a = 1530.50(18) pm
c = 1074.28(17) pm
V = 2179.3(5) ·10^6 pm^3
Z = 3
trigonal $R\bar{3}$

Pt:
a = 1541.0(3) pm
c = 1078.33(18) pm
V = 2217.7(7)
Z = 3
trigonal $R\bar{3}$

Isolated {ZHo$_6$} clusters connected directly via iodide ligands or via HoI$_6$ entities.

7. Summary and prospects

Dy{(C₂)Dy₆}I₁₂

a = 1523.3(3) pm
c = 1064.9(3) pm
V = 2140.0(7) ·10⁶ pm³
Z = 3
trigonal $R\bar{3}$

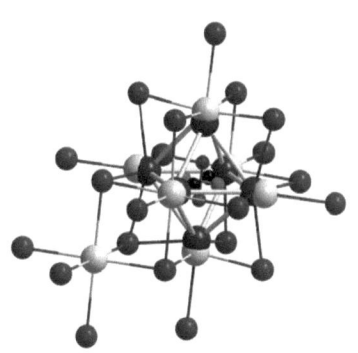

Isolated {(C$_2$)Dy$_6$} clusters connected directly via iodide ligands or via DyI$_6$ entities and encapsulating a disordered C$_2$ dumbbell.

Dy{CoDy₄.₅₃Y₁.₄₇}I₁₂

a = 1535.80(18) pm
c = 1078.83(18) pm
V = 2203.7(5) ·10⁶ pm³
Z = 3
trigonal $R\bar{3}$

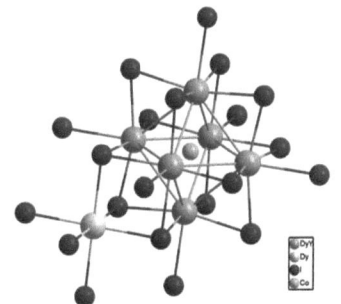

Isolated {CoDy$_{4.53}$Y$_{1.47}$} clusters connected directly via iodide ligands or via DyI$_6$ entities.

7. Summary and prospects

{(C$_2$)$_2$M$_{10}$}X$_{18}$ with M = Dy, X = Br and M = Er, X = I

Dy, Br:
a = 973.99(12) pm
b = 1633.98(15) pm
c = 1324.69(19) pm
β = 120.869(9)°
V = 1809.6(4) ·10^6 pm^3
Z = 2
monoclinic $P2_1/c$

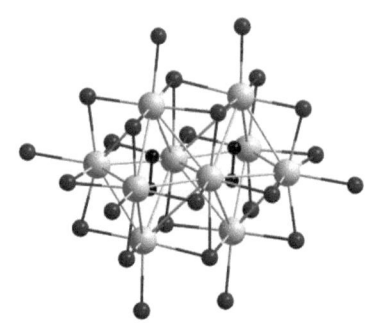

Er, I:
a = 1045.01(13) pm
b = 1707.22(17) pm
c = 1237.10(16) pm
β = 105.110(15)°
V = 2130.8(4) ·10^6 pm^3
Z = 2
monoclinic $P2_1/n$

Condensed, bi-octahedral {(C$_2$)$_2$M$_{10}$} units sharing a common edge and encapsulating C$_2$ dumbbells.

7. Summary and prospects

{Ru$_4$Ho$_{16}$}I$_{28}${Ho$_4$}

a = 1221.14(14) pm
V = 1820.9(4) ·10^6 pm^3
Z = 4
cubic $P\bar{4}3m$

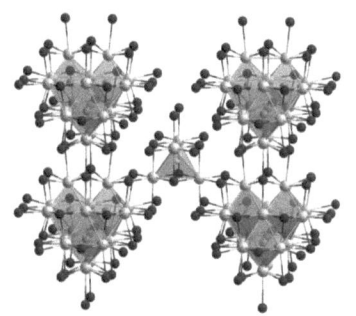

Tetrameric {Ru$_4$Ho$_{16}$}I$_{36}$ units connected directly via iodine ligands or empty {Ho$_4$}I$_8$ tetrahedra.

{(C$_2$)$_2$O$_2$Dy$_{14}$}I$_{24}$

a = 972.97(14) pm
b = 1033.03(13) pm
c = 1677.0(2) pm
α = 101.42(1)°
β = 92.72(1)°
γ = 112.75(1)°
V = 1509.3(3) ·10^6 pm^3
Z = 2
triclinic $P\bar{1}$

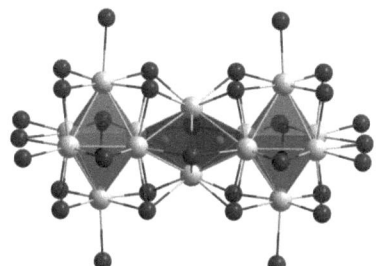

Tetrameric {M$_{14}$}I$_{32}$ clusters consisting of double tetrahedra encapsulating oxygen atoms flanked by octahedra containing C$_2$ dumbbells.

7. Summary and prospects

{(C₂)ODy₆}I₉

a = 2024.18(8) pm
c = 1299.21(4) pm
V = 4610.1(3) ·10⁶ pm³
Z = 8
hexagonal $P6/m$

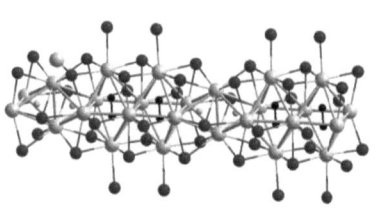

Chains consisting of double tetrahedra encapsulating oxygen atoms flanked by double octahedra containing C₂ dumbbells, i.e. a condensation product of the tetramer mentioned above.

{IrHo₃}I₃

a = 872.0(4) pm
b = 423.83(14) pm
c = 1211.2(6) pm
β = 94.99(4)°
V = 445.9(3) ·10⁶ pm³
Z = 2
monoclinic $P2_1/m$

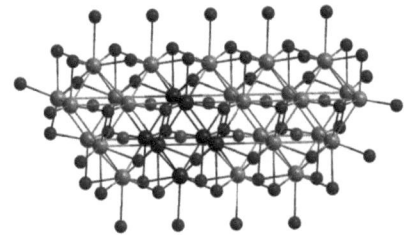

Double chains built by monocapped trigonal prismatic {Ho₇} clusters incorporating iridium atoms.

7. Summary and prospects

{(C$_2$)Er$_4$}I$_6$

a = 1354.71(14) pm
b = 1421.14(17) pm
c = 854.87(8) pm
V = 1645.8(3) ·10^6 pm^3
Z = 4
orthorhombic $P2_1/m$

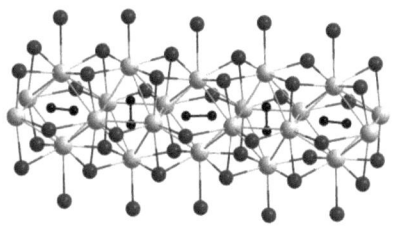

Chain consisting of edge-sharing {(C$_2$)Er$_6$} octahedra, alternately elongated and compressed.

References

[1] W. Klemm, H. Bommer. *Z. Anorg. Allg. Chem.*, 231:138, 1937.

[2] W. Döll, W. Klemm. *Z. Anorg. Allg. Chem.*, 241:239, 1939.

[3] H. Bärnighausen. *Proc. Rare Earth Res. Conf., 12th*, 1:404, 1976.

[4] U. Löchner, H. Bärnighausen, J.D. Corbett. *Inorg. Chem.*, 16:2134, 1977.

[5] G. Meyer. *Chem. Rev.*, 88:93, 1988.

[6] J.D. Martin, J.D. Corbett. *Angew. Chem, Int. Ed. Engl.*, 34:233, 1995.

[7] J.D. Corbett. *J. Chem. Soc., Dalton Trans.*, page 575, 1996.

[8] J.D. Corbett. *Inorg. Chem.*, 39:5178.

[9] A. Simon. *Angew. Chem.*, 100:163, 1988.

[10] A. Simon, H. Mattausch, M. Ryazanov, R.K. Kremer. *Z. Anorg. Allg. Chem.*, 632:919, 2006.

[11] S. Uhrlandt, G. Meyer. *Z. Anorg. Allg. Chem.*, 620:1872, 1994.

[12] L. Jongen, A.-V. Mudring, G. Meyer. *Angew. Chem.*, 118:1920, 2006.

[13] A. Palasyuk, I. Pantenburg, G. Meyer. *Acta Cryst.*, E62:i61, 2006.

[14] T. Hughbanks, J.D. Corbett. *Inorg. Chem.*, 27:2022, 1988.

[15] M. Payne, J.D. Corbett. *Inorg. Chem.*, 29:2246, 1990.

[16] H. Schäfer, H.G. Schnering. *Angew. Chem.*, 76:833, 1964.

References

[17] C. Felser, K. Ahn, R.K. Kremer, R.Seshadri, A. Simon. *J. Solid State Chem.*, 147:19, 1999.

[18] E. Warkentin, H. Bärnighausen. *Z. Anorg. Allg. Chem.*, 459:187, 1979.

[19] G. Meyer, A. Palasyuk. *Inorganic Chemistry in Focus III, ed. by G. Meyer, D. Naumann, L. Wesemann.* WILEY-VCH, 2006.

[20] H.P. Beck, M. Schuster. *J. Solid State Chem.*, 100:301, 1992.

[21] C. Hohnstedt. Dissertation, University of Hannover, 1993.

[22] A. Simon, Hj. Mattausch, G.J. Miller, W. Bauhofer, R.K. Kremer. *Handbook on the Physics and Chemistry of Rare Earths, ed. by K.A. Gscheidner, Jr. L. Eyring,* volume 15. Elsevier Science Publishers B.V., 1991.

[23] G. Meyer. *Z. Anorg. Allg. Chem.*, 634:2729, 2008.

[24] Hj. Mattausch, C. Hoch, A. Simon. *Z. Naturforsch.*, B62:148, 2007.

[25] A. Altomare, G. Cascarano, C. Giacovazzo, A. Gualardi. *J. Appl. Cryst.*, 26:343, 1993.

[26] G. Sheldrick. *SHELXS-97, A Program for Crystal Structure Solution.* University of Göttingen, 1997.

[27] G. Sheldrick. *SHELXL-97, A Program for Crystal Structure Refinement.* University of Göttingen, 1997.

[28] *STOE X-RED Vers. 1.07, Date Reduction for IPDS.* STOE CIE Darmstadt, 1996.

[29] *STOE X-SHAPE Vers. 1.01, Crystal Optimisation for Absorption Correction.* STOE CIE Darmstadt, 1996.

[30] L.R. Morss, G. Meyer. *Synthesis of Lanthanide and Actinide Compounds.* Kluwer Academic Publishers, Norwell, 1991.

[31] L.J. Farrugia. *J. Appl. Cryst.*, 32:837, 1999.

[32] L.J. Farrugia. *Platon00.* University of Glasgow, 2000.

[33] *STOE WinXPow Vers. 1.07.* STOE CIE Darmstadt, 2000.

References

[34] *X-AREA Vers. 1.16.* STOE CIE Darmstadt, 2002.

[35] *DIAMOND Vers. 3.0a, Crystal Structure Visualization.* Crystal Impact GbR Bonn, 2004.

[36] *Origin 7.0.* OriginLab Corporation, Northhampton, 2002.

[37] O.K. Andersen. *Phys. Rev. B: Condens. Matter Mater. Phys*, 12:3060, 1975.

[38] L.H. Skriver. *The LMTO Method.* Springer, Berlin, Germany, 1984.

[39] L. Hedin, B.I. Lundqvist. *J. Phys. Chem. C*, 4:2064, 1971.

[40] W.R.L. Lambrecht, O.K. Andersen. *Phys. Rev. B: Condens. Matter Mater. Phys*, 34:2439, 1986.

[41] P.E. Blöchl, O. Jepsen, O.K. Andersen. *Phys. Rev. B: Condens. Matter Mater. Phys*, 49:16223, 1994.

[42] R. Dronskowski, P.E. Blöchl. *J. Phys. Chem.*, 97:8617, 1993.

[43] T. Hughbanks, J.D. Corbett. *Inorg. Chem.*, 28:631, 1989.

[44] D.S. Dudis, J.D. Corbett. *Inorg. Chem.*, 26:1933, 1987.

[45] Y. Park, J.D. Corbett. *Inorg. Chem.*, 33:1705, 1994.

[46] G. Meyer, M.S. Wickleder. *Handbook on the Physics and Chemistry of Rare Earths, ed. by K.A. Gscheidner, Jr. & L. Eyring*, volume 28. Elsevier, Amsterdam, 2000.

[47] R. Wiglusz, I. Pantenburg, G. Meyer. *Z. Anorg. Allg. Chem.*, 633:1317, 2007.

[48] S. Satpathy, O.K. Andersen. *Inorg. Chem.*, 24:2604, 1985.

[49] C. Zheng, O. Oeckler, Hj. Mattausch, A. Simon. *Z. Anorg. Allg. Chem.*, 627:2151, 2001.

[50] G.J. Miller, J.K. Burdett, C. Schwarz, A. Simon. *Inorg. Chem.*, 25:4437, 1986.

[51] J.K. Burdett, G.J. Miller. *J. Am. Chem. Soc.*, 109:4092, 1987.

[52] G.-X. Xu, J. Ren. *J. Lanthanide Actinide Res.*, 2:67, 1987.

[53] S. Gao, G.-X. Xu, L.-M. Li. *Inorg. Chem.*, 31:4829, 1992.

References

[54] L.E. Sweet, L.E. Roy, F. Meng, T. Hughbanks. *J. Am. Chem. Soc.*, 128:10193, 2006.

[55] Hj. Mattausch, C. Zheng, L. Kienle, A. Simon. *Z. Anorg. Allg. Chem.*, 630:2367, 2004.

[56] M. Lukachuk, L. Kienle, C. Zheng, Hj. Mattausch, A. Simon. *Inorg. Chem.*, 47:4656, 2008.

[57] E. Warkentin, R. Masse, A. Simon. *Z. Anorg. Allg. Chem.*, 491:323, 1982.

[58] Hj. Mattausch, E. Warkentin, O. Oeckler, A. Simon. *Z. Anorg. Allg. Chem.*, 626:2117, 2000.

[59] Hj. Mattausch, A. Simon, L. Kienle, C. Hoch, C. Zheng, R.K. Kremer. *Z. Anorg. Allg. Chem.*, 632:1661, 2006.

[60] M. Brühmann. Diplomarbeit, University of Cologne, 2009.

[61] S. Zimmermann. Diplomarbeit, University of Cologne, 2006.

[62] S.J. Steinwand, J.D. Corbett. *Inorg. Chem.*, 35:7056, 1996.

[63] M.W. Payne, M. Ebihara, J.D. Corbett. *Angew. Chem. Int. Ed. Engl.*, 119:856, 1991.

[64] S. Zimmermann. Dissertation, University of Cologne, 2009.

[65] S.J. Steinwand, J.D. Corbett, J.D. Martin. *Inorg. Chem.*, 36:6413, 1997.

[66] M. Ebihara, J.D. Martin, J.D. Corbett. *Inorg. Chem.*, 33:2079, 1994.

[67] E.A. Owen, L. Pickup, J.O. Roberts. *Z. Kirstollogr.*, 91:70, 1935.

[68] E. Warkentin, H. Bärnighausen. *European Crystallographic Meeting*, 3:354, 1976.

[69] Hj. Mattausch, H. Borrmann, R. Eger, R.K. Kremer, A. Simon. *Z. Naturforsch.*, 50b:931, 1995.

[70] F. Steffen, G. Meyer. *Z. Naturforsch.*, 50:1570, 1995.

[71] M.W. Payne, P.K. Dorhout, J.D. Corbett. *Inorg. Chem.*, 30:1467, 1991.

[72] M.W. Payne, P.K. Dorhout, S.J. Kim, T.R. Hughbanks, J.D. Corbett. *Inorg. Chem.*, 31:1389, 1992.

References

[73] N. Herzmann, A.-V. Mudring, G. Meyer. *Inorg. Chem.*, 47:7954, 2008.

[74] H. Mattfeld, K. Krr, G. Meyer. *Z. anorg. allg. Chem.*, 619:1384, 1993.

[75] Hj. Mattausch, H. Borrmann, R. Eger, R.K. Kremer, A. Simon. *Z. anorg. allg. Chem.*, 620:1889, 1994.

[76] S.-J. Hwu, J.D. Corbett, K.R. Poeppelmeier. *J. Solid State Chem.*, 57:43, 1985.

[77] S.M. Kauzlarich, T. Hughbanks, J.D. Corbett, P. Klavins, R.N. Shelton. *Inorg. Chem.*, 27:1791, 1988.

[78] K. Berroth, A. Simon. *J. Less-Common Met.*, 76:41, 1980.

[79] C.C. Torardi, R.E. McCarley. *J. Am. Chem. Soc.*, 101:3963, 1979.

[80] P.O. Jeitschko, Hj. Mattausch, A. Simon. *Z. Anorg. Allg. Chem.*, 623:1815, 1997.

[81] J.D. Corbett. *personal communication.*

A. Appendix

Table A.1.: Atomic coordinates and equivalent displacement factors U_{eq} (10^{-4} pm^2) in Ho{FeHo$_6$}I$_{12}$

Atom	x/a	y/b	z/c	U_{eq}
Ho1	1.15739(5)	0.04355(5)	0.63807(6)	0.0153(2)
Ho2	1.00000	0.00000	1.00000	0.0213(4)
I3	1.05135(6)	-0.13025(7)	0.83941(7)	0.0190(2)
I4	1.31674(7)	0.23705(7)	0.50663(8)	0.0242(3)
Fe5	1.00000	0.00000	0.50000	0.0132(9)

Table A.2.: Anisotropic displacement factors U_{ij} (10^{-4} pm^2) in Ho{FeHo$_6$}I$_{12}$

Atom	U_{11}	U_{22}	U_{33}	U_{12}	U_{13}	U_{23}
Ho1	0.0155(3)	0.0161(3)	0.0147(3)	0.0083(3)	-0.0004(2)	-0.0002(2)
Ho2	0.0223(5)	0.0223(5)	0.0194(7)	0.0111(3)	0.00000	0.00000
I3	0.0202(5)	0.0195(5)	0.0183(4)	0.0106(4)	-0.0006(3)	0.0010(3)
I4	0.0167(5)	0.0233(5)	0.0260(5)	0.0050(4)	-0.0030(3)	0.0060(3)

Table A.3.: Atomic coordinates and equivalent displacement factors U_{eq} (10^{-4} pm^2) in Ho{NiHo$_6$}I$_{12}$

Atom	x/a	y/b	z/c	U_{eq}
Ho1	0.82465(2)	0.44785(2)	-0.52672(3)	0.0159(1)
Ho2	0.66670	0.33330	-0.16670	0.0277(2)
I4	0.71896(4)	0.51892(4)	-0.33848(5)	0.0221(2)
I5	0.98441(4)	0.41572(4)	-0.66061(5)	0.0251(2)
Ni3	0.66670	0.33330	-0.66670	0.0125(4)

Table A.4.: Anisotropic displacement factors U_{ij} (10^{-4} pm^2) in Ni{FeHo$_6$}I$_{12}$

Atom	U_{11}	U_{22}	U_{33}	U_{12}	U_{13}	U_{23}
Ho1	0.01342(18)	0.01456(19)	0.0186(2)	0.00612(13)	-0.00137(11)	-0.00088(10)
Ho2	0.0302(3)	0.0302(3)	0.0226(4)	0.01511(15)	0.00000	0.00000
I4	0.0217(3)	0.0208(3)	0.0240(3)	0.01068(19)	-0.00145(16)	-0.00437(17)
I5	0.0161(2)	0.0297(3)	0.0291(3)	0.0111(2)	-0.00389(17)	-0.00964(19)

A. Appendix

Table A.5.: Atomic coordinates and equivalent displacement factors U_{eq} (10^{-4} pm^2) in Ho{PtHo$_6$}I$_{12}$

Atom	x/a	y/b	z/c	U_{eq}
Ho1	1.16126(7)	0.04464(7)	0.64383(8)	0.0172(2)
Ho2	1.00000	0.00000	1.00000	0.0316(6)
I3	1.05255(11)	-0.13264(11)	0.83011(15)	0.0242(3)
I4	1.31737(11)	0.23536(13)	0.50600(16)	0.0268(4)
Pt5	1.00000	0.00000	0.50000	0.0125(4)

Table A.6.: Anisotropic displacement factors U_{ij} (10^{-4} pm^2) in Ho{PtHo$_6$}I$_{12}$

Atom	U_{11}	U_{22}	U_{33}	U_{12}	U_{13}	U_{23}
Ho1	0.0151(4)	0.0191(4)	0.0182(4)	0.0092(4)	-0.0029(3)	-0.0012(3)
Ho2	0.0349(9)	0.0349(9)	0.0251(15)	0.0174(5)	0.00000	0.00000
I3	0.0251(6)	0.0234(6)	0.0238(7)	0.0119(5)	-0.0015(5)	0.0032(5)
I4	0.0164(6)	0.0268(7)	0.0305(10)	0.0058(5)	-0.0039(5)	0.0066(5)
Pt5	0.0119(5)	0.0119(5)	0.0136(9)	0.0059(2)	0.00000	0.00000

Table A.7.: Atomic coordinates and equivalent displacement factors U_{eq} (10^{-4} pm^2) in Dy{(C$_2$)Dy$_6$}I$_{12}$

Atom	x/a	y/b	z/c	S.O.F.	U_{eq}
Dy1A	0.9593(2)	-0.1434(7)	0.1268(6)	0.57	0.0196(9)
Dy1B	0.9528(4)	-0.1711(13)	0.1516(13)	0.43	0.0262(16)
I2	1.08093(7)	-0.23313(6)	-0.00199(8)		0.0332(3)
Dy3	1.00000	0.00000	0.50000		0.0411(4)
I4	1.13244(6)	-0.05244(6)	0.32362(8)		0.0288(2)
C	0.9884(17)	-0.044(2)	0.041(2)	0.33	0.004(4)

Table A.8.: Anisotropic displacement factors U_{ij} (10^{-4} pm^2) in Dy{(C$_2$)Dy$_6$}I$_{12}$

Atom	U_{11}	U_{22}	U_{33}	U_{12}	U_{13}	U_{23}
Dy1A	0.0223(6)	0.0156(17)	0.0208(11)	0.0094(7)	-0.0005(5)	0.0036(11)
Dy1B	0.0254(10)	0.025(3)	0.030(2)	0.0134(13)	-0.0022(10)	-0.002(2)
I2	0.0363(5)	0.0298(5)	0.0372(5)	0.0193(4)	0.0050(3)	0.0063(3)
Dy3	0.0459(6)	0.0459(6)	0.0314(8)	0.0230(3)	0.00000	0.00000
I4	0.0277(4)	0.0288(4)	0.0306(4)	0.0148(3)	-0.0025(3)	0.0007(3)

Table A.9.: Atomic coordinates and equivalent displacement factors U_{eq} (10^{-4} pm^2) in Dy$\{CoDy_{4.53}Y_{1.47}\}I_{12}$

Atom	x/a	y/b	z/c	S.O.F.	U_{eq}
Y1	0.88493(6)	-0.15900(6)	0.14158(7)	0.211	0.0202(3)
Dy1	0.88493(6)	-0.15900(6)	0.14158(7)	0.789	0.0202(3)
Dy2	1.00000	0.00000	0.50000		0.0360(5)
I3	0.91768(8)	-0.31725(8)	0.00516(8)		0.0272(3)
I4	0.81444(7)	-0.05253(7)	0.32778(8)		0.0245(3)
Co5	1.00000	0.00000	0.00000		0.0151(15)

Table A.10.: Anisotropic displacement factors U_{ij} (10^{-4} pm^2) in Dy$\{CoDy_{4.53}Y_{1.47}\}I_{12}$

Symbol	U_{11}	U_{22}	U_{33}	U_{12}	U_{13}	U_{23}
Y1	0.0202(4)	0.0195(4)	0.0195(4)	0.0088(3)	0.0010(3)	0.0014(3)
Dy1	0.0202(4)	0.0195(4)	0.0195(4)	0.0088(3)	0.0010(3)	0.0014(3)
Dy2	0.0417(8)	0.0417(8)	0.0246(9)	0.0209(4)	0.00000	0.00000
I3	0.0339(6)	0.0204(5)	0.0267(6)	0.0131(4)	0.0091(4)	0.0033(3)
I4	0.0260(6)	0.0260(6)	0.0218(5)	0.0133(4)	0.0054(3)	0.0020(3)
Co5	0.0159(18)	0.0159(18)	0.014(2)	0.0080(9)	0.00000	0.00000

A. Appendix

Table A.11.: Atomic coordinates and equivalent displacement factors U_{eq} (10^{-4} pm^2) in $\{(C_2)_2Dy_{10}\}Br_{18}$

Atom	x/a	y/b	z/c	U_{eq}
Dy1	0.72140(7)	-0.18631(3)	0.55464(5)	0.0164(1)
Dy2	0.86572(6)	-0.05763(3)	0.81405(4)	0.0158(1)
Dy3	0.49120(6)	0.16312(3)	0.27712(5)	0.0165(1)
Dy4	0.66436(6)	0.04537(3)	0.53869(5)	0.0167(1)
Dy6	0.49942(6)	-0.05623(3)	0.23673(5)	0.0164(1)
Br11	0.79002(15)	0.17718(7)	0.46836(11)	0.0267(3)
Br12	0.18295(15)	-0.05866(7)	0.03920(11)	0.0261(3)
Br13	0.81121(14)	-0.06564(6)	0.44345(10)	0.0199(2)
Br14	0.19236(15)	0.18349(7)	0.05984(11)	0.0270(3)
Br15	0.60176(17)	-0.19122(6)	0.14246(12)	0.0282(3)
Br16	0.59583(16)	0.06929(7)	0.14910(12)	0.0273(3)
Br17	1.01722(16)	-0.18152(7)	0.75532(12)	0.0274(3)
Br19	0.97720(15)	0.06689(7)	0.73923(11)	0.0265(3)
Br20	0.58632(14)	0.19733(6)	0.66855(10)	0.0198(2)
C1	0.6314(13)	-0.0891(5)	0.6212(9)	0.0118(18)
C2	0.4262(13)	0.0309(6)	0.3277(9)	0.0144(19)

Table A.12.: Anisotropic displacement factors U_{ij} (10^{-4} pm^2) in $\{(C_2)_2Dy_{10}\}Br_{18}$

Atom	U_{11}	U_{22}	U_{33}	U_{12}	U_{13}	U_{23}
Dy1	0.0185(3)	0.0131(2)	0.0193(3)	0.00139(17)	0.0109(2)	-0.00086(17)
Dy2	0.0133(3)	0.0157(2)	0.0161(3)	-0.00058(17)	0.0058(2)	0.00070(17)
Dy3	0.0170(3)	0.0136(2)	0.0190(3)	-0.00068(17)	0.0094(2)	0.00148(17)
Dy4	0.0136(3)	0.0162(2)	0.0177(3)	-0.00185(17)	0.0061(2)	0.00218(17)
Dy6	0.0162(3)	0.0138(2)	0.0191(3)	0.00202(17)	0.0089(2)	-0.00146(17)
Br11	0.0196(6)	0.0273(5)	0.0261(6)	-0.0087(4)	0.0066(5)	0.0045(4)
Br12	0.0170(6)	0.0321(6)	0.0209(6)	0.0031(4)	0.0037(5)	-0.0070(5)
Br13	0.0161(6)	0.0213(5)	0.0235(6)	0.0005(4)	0.0110(5)	0.0020(4)
Br14	0.0214(7)	0.0287(5)	0.0247(6)	-0.0059(4)	0.0073(5)	0.0103(4)
Br15	0.0435(8)	0.0148(5)	0.0400(7)	0.0018(4)	0.0313(7)	-0.0017(4)
Br16	0.0361(7)	0.0225(5)	0.0354(7)	-0.0012(5)	0.0271(6)	0.0002(5)
Br17	0.0191(6)	0.0298(5)	0.0281(7)	0.0066(4)	0.0083(5)	-0.0031(4)
Br19	0.0182(6)	0.0291(5)	0.0248(6)	-0.0074(4)	0.0057(5)	0.0056(4)
Br20	0.0189(6)	0.0163(4)	0.0235(6)	-0.0005(4)	0.0104(5)	-0.0009(4)

Table A.13.: Atomic coordinates and equivalent displacement factors U_{eq} (10^{-4} pm^2) in $\{(C_2)_2Er_{10}\}I_{18}$

Atom	x/a	y/b	z/c	U_{eq}
Er1	1.24414(8)	-0.05453(5)	0.75428(6)	0.0084(2)
Er2	0.85173(8)	0.18127(5)	0.55373(6)	0.0087(2)
Er3	0.95776(8)	0.05751(5)	0.80379(6)	0.0098(2)
Er4	1.20443(8)	0.15658(5)	0.71781(6)	0.0098(2)
Er5	1.11542(9)	0.04492(5)	0.46153(7)	0.0135(2)
I6	1.36426(11)	-0.06696(7)	0.55619(9)	0.0122(2)
I7	1.24060(14)	0.06552(8)	0.25925(11)	0.0210(3)
I8	1.44475(12)	0.06463(7)	0.8481(1)	0.0177(3)
I9	1.13922(13)	0.18399(8)	0.94195(10)	0.0180(3)
I10	1.32683(14)	0.17893(8)	0.53457(11)	0.0209(3)
I11	0.74430(13)	0.18413(8)	0.75023(10)	0.0190(3)
I12	0.93159(12)	0.19847(7)	0.33182(10)	0.0121(2)
I13	1.14365(13)	-0.06096(8)	0.96066(10)	0.0182(3)
I14	1.04656(13)	0.31059(7)	0.64168(10)	0.0182(3)
C1	0.9964(17)	0.0897(10)	0.6177(13)	0.008(3)
C2	1.0934(16)	0.0283(10)	0.6700(13)	0.007(3)

Table A.14.: Anisotropic displacement factors U_{ij} (10^{-4} pm^2) in $\{(C_2)_2Er_{10}\}I_{18}$

Atom	U_{11}	U_{22}	U_{33}	U_{12}	U_{13}	U_{23}
Er1	0.0117(4)	0.0041(4)	0.0101(4)	0.0029(3)	0.0041(3)	0.0014(3)
Er2	0.0115(4)	0.0045(4)	0.0104(4)	0.0018(3)	0.0033(3)	0.0011(3)
Er3	0.0132(4)	0.0070(4)	0.0102(4)	-0.0007(3)	0.0048(3)	-0.0004(3)
Er4	0.0129(4)	0.0067(4)	0.0107(4)	-0.0011(3)	0.0049(3)	0.0000(3)
Er5	0.0171(4)	0.0109(4)	0.0141(4)	-0.0025(4)	0.0072(3)	-0.0029(3)
I6	0.0118(5)	0.0111(6)	0.0148(5)	-0.0010(5)	0.0054(5)	-0.0025(4)
I7	0.0245(7)	0.0192(7)	0.0221(6)	-0.0090(6)	0.0110(6)	-0.0020(5)
I8	0.0157(6)	0.0112(6)	0.0230(7)	-0.0009(5)	-0.0008(5)	-0.0007(5)
I9	0.0238(6)	0.0157(6)	0.0168(6)	-0.0118(6)	0.0093(5)	-0.0079(5)
I10	0.0260(6)	0.0202(7)	0.0207(6)	-0.0065(6)	0.0136(5)	-0.0016(5)
I11	0.0233(6)	0.0183(7)	0.0196(6)	0.0077(6)	0.0132(5)	0.0011(5)
I12	0.0168(5)	0.0079(5)	0.0133(5)	-0.0006(5)	0.0068(4)	0.0014(4)
I13	0.0239(6)	0.0201(7)	0.0143(6)	0.0064(6)	0.0114(5)	0.0063(5)
I14	0.0202(6)	0.0060(6)	0.0245(7)	-0.0029(5)	-0.0010(5)	-0.0005(4)

A. Appendix

Table A.15.: Atomic coordinates and equivalent displacement factors U_{eq} (10^{-4} pm^2) in {Ru$_4$Ho$_{16}$}I$_{28}${Ho$_4$}

Atom	x/a	y/b	z/c	U_{eq}
Ho1	0.12651(13)	0.12651(13)	0.12651(13)	0.0130(6)
Ho2	0.31997(11)	0.11183(11)	-0.11183(11)	0.0152(3)
Ho3	0.37103(17)	0.37103(17)	-0.37103(17)	0.0355(9)
I4	0.3963(2)	0.13238(16)	0.13238(16)	0.0186(5)
I5	0.36501(16)	0.1142(3)	-0.36501(16)	0.0217(6)
I6	0.6279(2)	0.3721(2)	-0.3721(2)	0.0277(10)
Ru1	0.1002(2)	0.1002(2)	-0.1002(2)	0.0130(9)

Table A.16.: Anisotropic displacement factors U_{ij} (10^{-4} pm^2) in {Ru$_4$Ho$_{16}$}I$_{28}${Ho$_4$}

Atom	U_{11}	U_{22}	U_{33}	U_{12}	U_{13}	U_{23}
Ho1	0.0130(6)	0.0130(6)	0.0130(6)	0.0011(8)	0.0011(8)	0.0011(8)
Ho2	0.0133(8)	0.0162(5)	0.0162(5)	0.0015(5)	-0.0015(5)	-0.0032(9)
Ho3	0.0355(9)	0.0355(9)	0.0355(9)	-0.0033(10)	0.0033(10)	0.0033(10)
I4	0.0152(13)	0.0203(8)	0.0203(8)	-0.0030(8)	-0.0030(8)	-0.0003(11)
I5	0.0184(9)	0.0283(16)	0.0184(9)	0.0006(10)	0.0007(10)	-0.0006(10)
I6	0.0277(10)	0.0277(10)	0.0277(10)	0.0029(13)	-0.0029(13)	0.0029(13)
Ru7	0.0130(9)	0.0130(9)	0.0130(9)	0.0016(14)	-0.0016(14)	-0.0016(14)

Table A.17.: Atomic coordinates and equivalent displacement factors U_{eq} (10^{-4} pm^2) in $\{(C_2)_2O_2Dy_{14}\}I_{24}$

Atom	x/a	y/b	z/c	U_{eq}
Dy1	0.33744(8)	0.45717(8)	0.11200(4)	0.0223(2)
Dy2	0.06054(8)	0.43229(8)	0.26401(5)	0.0235(2)
Dy3	0.12895(8)	0.11147(8)	0.14223(5)	0.0223(2)
Dy4	0.54954(8)	0.24910(8)	0.18965(5)	0.0228(2)
Dy5	0.27867(8)	0.23735(8)	0.34940(5)	0.0236(2)
Dy10	0.48111(8)	0.56997(8)	0.31943(5)	0.0227(2)
Dy11	-0.35579(9)	0.44995(10)	0.47997(5)	0.0296(2)
I1	0.33894(13)	-0.03256(12)	0.08614(8)	0.0318(2)
I2	0.04486(14)	-0.06310(12)	0.26972(8)	0.0345(3)
I3	0.56535(13)	0.75450(12)	0.19662(7)	0.0326(2)
I4	0.09086(13)	0.57230(13)	0.12380(8)	0.0327(2)
I5	0.54473(13)	0.13960(12)	0.35052(7)	0.0301(2)
I6	-0.20921(11)	0.53355(12)	0.31216(7)	0.0284(2)
I7	0.27041(12)	0.71323(12)	0.39081(7)	0.0303(2)
I8	-0.17795(12)	0.14985(12)	0.16271(8)	0.0317(2)
I9	0.38877(13)	0.60146(13)	-0.04848(7)	0.0320(2)
I10	0.29650(14)	0.17817(13)	0.53055(7)	0.0357(3)
I11	0.12315(13)	0.19765(12)	-0.02137(7)	0.0304(2)
I20	0.02449(12)	0.31908(13)	0.42583(7)	0.0313(2)
O1	0.4437(15)	0.4555(14)	0.4171(8)	0.034(3)
C1	0.3611(18)	0.3169(17)	0.2157(10)	0.024(3)
C2	0.245(2)	0.3613(19)	0.2361(11)	0.030(3)

A. Appendix

Table A.18.: Anisotropic displacement factors U_{ij} (10^{-4} pm^2) in $\{(C_2)_2O_2Dy_{14}\}I_{24}$

Atom	U_{11}	U_{22}	U_{33}	U_{12}	U_{13}	U_{23}
Dy1	0.0209(3)	0.0259(3)	0.0207(3)	0.0100(3)	0.0030(3)	0.0059(3)
Dy2	0.0199(3)	0.0305(4)	0.0228(3)	0.0136(3)	0.0038(3)	0.0046(3)
Dy3	0.0193(3)	0.0249(3)	0.0217(3)	0.0089(3)	0.0013(3)	0.0035(3)
Dy4	0.0187(3)	0.0291(3)	0.0234(3)	0.0128(3)	0.0037(3)	0.0057(3)
Dy5	0.0211(3)	0.0247(3)	0.0233(3)	0.0085(3)	0.0020(3)	0.0040(3)
Dy10	0.0208(3)	0.0269(3)	0.0218(3)	0.0104(3)	0.0031(3)	0.0071(3)
Dy11	0.0276(4)	0.0449(5)	0.0195(3)	0.0212(3)	0.0007(3)	0.0018(3)
I1	0.0282(5)	0.0300(5)	0.0363(6)	0.0146(4)	0.0022(4)	0.0003(4)
I2	0.0348(6)	0.0289(5)	0.0321(6)	0.0049(4)	0.0013(5)	0.0079(4)
I3	0.0337(6)	0.0290(5)	0.0292(5)	0.0054(4)	0.0008(4)	0.0093(4)
I4	0.0321(6)	0.0409(6)	0.0351(6)	0.0222(5)	0.0076(5)	0.0148(5)
I5	0.0324(6)	0.0353(5)	0.0296(5)	0.0199(5)	0.0062(4)	0.0099(4)
I6	0.0209(5)	0.0334(5)	0.0318(5)	0.0136(4)	0.0047(4)	0.0040(4)
I7	0.0275(5)	0.0316(5)	0.0322(6)	0.0150(4)	0.0024(4)	0.0024(4)
I8	0.0206(5)	0.0322(5)	0.0400(6)	0.0125(4)	0.0022(4)	0.0000(4)
I9	0.0319(6)	0.0433(6)	0.0309(6)	0.0216(5)	0.0122(5)	0.0160(5)
I10	0.0383(6)	0.0369(6)	0.0260(5)	0.0101(5)	-0.0004(4)	0.0064(4)
I11	0.0318(5)	0.0295(5)	0.0239(5)	0.0076(4)	-0.0012(4)	0.0040(4)
I20	0.0265(5)	0.0418(6)	0.0292(5)	0.0160(5)	0.0083(4)	0.0104(5)

Table A.19.: Atomic coordinates and equivalent displacement factors U_{eq} (10^{-4} pm^2) in $\{(C_2)ODy_6\}I_9$

Atom	x/a	y/b	z/c	S.O.F.	U_{eq}
Dy1	0.12072(3)	0.46127(3)	0.34259(4)		0.0252(1)
Dy2	0.40742(3)	0.47350(3)	0.21988(4)		0.0265(1)
Dy3	0.41098(4)	0.47215(4)	0.50000		0.0229(2)
Dy11	0.52210(6)	0.59192(5)	0.00000		0.0336(2)
Dy30	0.15265(9)	0.13255(9)	0.15758(12)	0.33	0.0226(3)
Dy31	0.09166(13)	0.03624(13)	0.00000	0.33	0.0217(5)
Dy32	0.05764(9)	0.09512(9)	0.27982(12)	0.33	0.0218(3)
Dy33	0.01256(15)	0.08904(14)	0.50000	0.33	0.0266(5)
I10	0.26534(4)	0.42852(4)	0.34868(6)		0.0279(2)
I14	0.33231(7)	0.28892(7)	0.50000		0.0288(2)
I15	0.41662(6)	0.63010(6)	0.50000		0.0251(2)
I18	0.42927(5)	0.63377(5)	0.16538(6)		0.0336(2)
I21	0.34767(6)	0.30024(6)	0.16583(6)		0.0409(2)
I29	0.31855(7)	0.44709(8)	0.00000		0.0367(3)
I33	0.12481(9)	0.20568(7)	0.33431(10)	0.67	0.0458(4)
I34	0.22264(11)	0.0883(2)	0.00000	0.67	0.0687(9)
I35	0.23074(12)	0.09166(13)	-0.14735(17)	0.33	0.0261(4)
I36	0.10859(19)	0.18070(18)	0.50000	0.33	0.0268(7)
O1	0.50000	0.50000	0.1104(11)		0.040(3)
C1	0.0298(6)	0.4903(6)	0.3536(9)		0.024(2)
O2	0.00000	0.00000	0.3918(16)		0.028(4)
C2	0.017(2)	0.0434(13)	0.147(2)	0.33	0.013(6)

Table A.20.: Anisotropic displacement factors U_{ij} (10^{-4} pm^2) in $\{(C_2)ODy_6\}I_9$

Atom	U_{11}	U_{22}	U_{33}	U_{12}	U_{13}	U_{23}
Dy1	0.0247(3)	0.0327(3)	0.0172(2)	0.0135(2)	0.0006(2)	-0.0012(2)
Dy2	0.0309(3)	0.0378(3)	0.0170(2)	0.0218(3)	0.0003(2)	-0.0004(2)
Dy3	0.0281(4)	0.0293(4)	0.0151(3)	0.0173(3)	0.00000	0.00000
Dy11	0.0521(5)	0.0394(5)	0.0170(4)	0.0286(4)	0.00000	0.00000
I10	0.0294(4)	0.0374(4)	0.0206(3)	0.0195(3)	0.0009(3)	0.0000(3)
I14	0.0355(6)	0.0362(6)	0.0192(5)	0.0213(5)	0.00000	0.00000
I15	0.0288(5)	0.0304(5)	0.0211(5)	0.0185(4)	0.00000	0.00000
I18	0.0470(5)	0.0451(5)	0.0228(4)	0.0336(4)	-0.0053(3)	-0.0026(3)
I21	0.0544(5)	0.0535(5)	0.0233(4)	0.0333(5)	-0.0099(4)	-0.0052(4)
I29	0.0386(6)	0.0571(8)	0.0215(5)	0.0292(6)	0.00000	0.00000
Dy30	0.0227(7)	0.0236(7)	0.0179(7)	0.0089(6)	0.0000(6)	-0.0001(6)
Dy31	0.0224(10)	0.0262(11)	0.0152(10)	0.0111(9)	0.00000	0.00000
Dy32	0.0233(8)	0.0240(7)	0.0173(7)	0.0113(6)	0.0003(6)	0.0001(6)
Dy33	0.0404(14)	0.0275(11)	0.0153(10)	0.0196(11)	0.00000	0.00000
I33	0.0708(9)	0.0239(6)	0.0214(6)	0.0076(6)	-0.0028(6)	-0.0021(5)
I34	0.025(1)	0.137(3)	0.0176(9)	0.0209(12)	0.00000	0.00000
I35	0.0236(10)	0.0310(11)	0.0225(11)	0.0128(9)	-0.0001(9)	0.0004(9)
I36	0.0295(16)	0.0289(16)	0.0208(15)	0.0136(14)	0.00000	0.00000

Table A.21.: Atomic coordinates and equivalent displacement factors U_{eq} (10^{-4} pm^2) in $\{IrHo_3\}I_3$

Atom	x/a	y/b	z/c	U_{eq}
Ho1	-0.1459(7)	-0.25000	0.1315(6)	0.0142(15)
Ho2	0.0994(8)	0.25000	0.3199(7)	0.0183(16)
Ho3	0.2526(8)	-0.25000	0.1024(6)	0.0172(16)
I1	0.4312(11)	-0.25000	0.8612(9)	0.021(2)
I2	0.3616(11)	-0.25000	0.3670(9)	0.020(2)
I3	-0.1463(11)	-0.25000	0.3941(8)	0.017(2)
Ir1	0.0469(6)	0.25000	0.0919(5)	0.0135(13)

Table A.22.: Anisotropic displacement factors U_{ij} (10^{-4} pm^2) in $\{IrHo_3\}I_3$

Atom	U_{11}	U_{22}	U_{33}	U_{12}	U_{13}	U_{23}
Ho1	0.011(3)	0.008(4)	0.024(3)	0.00000	0.004(2)	0.00000
Ho2	0.013(3)	0.010(4)	0.033(4)	0.00000	0.003(3)	0.00000
Ho3	0.012(3)	0.017(4)	0.023(3)	0.00000	-0.002(3)	0.00000
I1	0.016(4)	0.017(6)	0.032(6)	0.00000	0.004(4)	0.00000
I2	0.010(4)	0.016(5)	0.032(5)	0.00000	-0.007(4)	0.00000
I3	0.014(4)	0.017(5)	0.019(4)	0.00000	0.004(4)	0.00000
Ir1	0.012(2)	0.011(3)	0.019(3)	0.00000	0.005(2)	0.00000

Table A.23.: Atomic coordinates and equivalent displacement factors U_{eq} (10^{-4} pm^2) in $\{(C_2)Er_4\}I_6$

Atom	x/a	y/b	z/c	S.O.F.	U_{eq}
Er1	-0.36491(8)	-0.34095(7)	0.50000		0.0469(3)
Er2A	-0.3950(2)	-0.3749(3)	0.00000	0.84	0.0482(7)
Er2B	-0.4236(15)	-0.4091(16)	0.00000	0.16	0.050(3)
Er3	-0.59506(6)	-0.42713(6)	0.29070(16)		0.0606(3)
I1	-0.48319(8)	-0.23839(8)	0.24820(16)		0.0507(3)
I2	-0.22137(15)	-0.16693(13)	0.50000		0.0586(5)
I3	-0.23556(8)	-0.42787(8)	0.24651(16)		0.0502(3)
I4	-0.24101(13)	-0.17993(13)	0.00000		0.0589(5)
C1	-0.4659(19)	-0.4587(19)	0.50000		0.052(6)
C2	-0.517(7)	-0.491(8)	0.089(7)	0.5	0.11(2)

Table A.24.: Anisotropic displacement factors U_{ij} (10^{-4} pm^2) in $\{(C_2)Er_4\}I_6$

Atom	U_{11}	U_{22}	U_{33}	U_{12}	U_{13}	U_{23}
Er1	0.0381(5)	0.0381(5)	0.0645(8)	-0.0017(4)	0.00000	0.00000
Er2A	0.0452(11)	0.0453(13)	0.0541(9)	0.0019(9)	0.00000	0.00000
Er2B	0.060(6)	0.052(6)	0.039(4)	-0.002(5)	0.00000	0.00000
Er3	0.0357(4)	0.0378(4)	0.1083(9)	0.0011(3)	0.0016(4)	0.0018(4)
I1	0.0444(5)	0.0415(5)	0.0663(8)	0.0006(4)	0.0005(5)	0.0001(5)
I2	0.0684(11)	0.0492(9)	0.0582(11)	-0.0032(8)	0.00000	0.00000
I3	0.0391(5)	0.0451(6)	0.0662(8)	-0.0007(4)	-0.0005(5)	-0.0026(5)
I4	0.0462(8)	0.0499(9)	0.0806(13)	-0.0117(7)	0.00000	0.00000

Die VDM Verlagsservicegesellschaft sucht für wissenschaftliche Verlage abgeschlossene und herausragende

Dissertationen, Habilitationen, Diplomarbeiten, Master Theses, Magisterarbeiten usw.

für die kostenlose Publikation als Fachbuch.

Sie verfügen über eine Arbeit, die hohen inhaltlichen und formalen Ansprüchen genügt, und haben Interesse an einer honorarvergüteten Publikation?

Dann senden Sie bitte erste Informationen über sich und Ihre Arbeit per Email an *info@vdm-vsg.de*.

Sie erhalten kurzfristig unser Feedback!

VDM Verlagsservicegesellschaft mbH
Dudweiler Landstr. 99
D - 66123 Saarbrücken

Telefon +49 681 3720 174
Fax +49 681 3720 1749

www.vdm-vsg.de

Die VDM Verlagsservicegesellschaft mbH vertritt

Printed by Books on Demand GmbH, Norderstedt / Germany